THE REVOLUTIONARY PHENOTYPE

J. -F. Gariépy

For those who were unjustly taken against my will: R., J., É.

For those who were not: É.

Contents

Letter to the Reader

It has now been more than five years that I've been reflecting on the theory laid out in this book. Most employed scientists today do not have the leisure to spend that much time doing science. They spend most of their days writing grant proposals and asking the permission of ethics committees. Long-term thinking is not rewarded in academia, as a scientist must produce something "good" every year or so in order to remain competitive. The quality of scientific works, as a result, has greatly diminished.

I quickly realized that if I maintained employment in a university, I would be unable to put in the intellectual efforts this scientific theory deserved. When I discovered the first few elements of this theory, I resigned from my postdoctoral position at Duke University. I knew that anything I would spend my time on in the academic world would inevitably be less important than this book.

Along the way, I had to find survival strategies other than begging the state for money to fund my scientific project. It turned out to be quite fascinating. I started a YouTube career and I now make a living out of discussing science, news and politics on the Internet.

A theory as important as the one laid out in this book usually comes with a letter of introduction from a well-known scientist in the field. I did not solicit such letters. The truth needs no introduction. Furthermore, the theory contained in this book is important enough that, one day, any person interested in the fate of humanity will feel obligated to read and share it.

I do, however, have a commitment to honor through this letter. When I was starting out, one of the early members of my audience said he would send me $10 if I promised to write a letter about him as the introduction to my book. He went by a pseudonym, Abraham

Lincoln. I accepted his deal, and so here is your letter, Abraham, in a book I was able to write thanks to people like you.

Across the years, I have gathered millions of views on YouTube for my news and educational videos. The core of my audience has come to be a highly intellectual and engaging crowd. I feel at home among them even more than I used to in my monkey laboratory. They are some of the most critical and intelligent people I know.

So, I thank all of them for participating in this community and for having indirectly funded this important book. And I thank all of you for picking up this book. If you are holding this book in your hands, the fate of humanity depends on you.

Prologue

How would you react if, in the first few lines of a book, the author told you the book you've just opened contains the complete description of how humanity will almost certainly destroy itself? Would you close the book and hope that no one else reads it? Or would you absolutely want to read it and spread the word about its contents to others?

I'm not asking these questions merely as a thought experiment. You are indeed holding a book which portends the end of DNA-based life on Earth. Most of the technology required to implement this carnage has already been developed and there are no international laws keeping anyone from using it.

Furthermore, there is very little hope that enough people will read this book and be motivated enough to keep this slaughter from happening. If anything, a naive part of humanity may very well read this book and be encouraged by it, joyfully welcoming humanity's upcoming extinction.

Hopefully, though, the information contained in this book will motivate humans to look at the path we have so naively engaged in. If the readers of this book act in concert, we may be able to stop humanity from committing a monumental mistake.

This book delves into a fascinating, yet very rare, type of biological events. During these events, groups of naive creatures fabricate other, better, life forms. Soon after they are born, the novel life forms end up annihilating the ancient ones that spawned them.

According to modern science, such events *have* occurred… repeatedly—as many as three times within the past few billion years on planet Earth alone. *Homo sapiens* are about to become the first earthly species to develop an awareness of them. Simultaneously, we may also become the first life form to acquire the requisite knowledge to prevent such events from recurring. To paraphrase George Santayana, *those life forms who cannot infer their forebear's past revolutions are condemned to repeat them.*

The biological events described in this book are nothing short of apocalyptic. Imagine a group of living creatures who casually decide to fabricate another life form, only to realize as they see their creation come alive that it will eventually outperform them and ultimately take over. This scenario is *not* discussed in the biology textbooks of your local state-funded university. Many teachers of biology prefer relegating such considerations to the realm of science fiction. However, in doing so, they deny current scientific advancements, for biology has now matured to a point where we can no longer treat this doomsday scenario as hypothetical.

It is a fact that DNA-based life was created by another life form. The ancient life form was called RNA. Sometime after the creation of DNA, the RNA organisms lost control over it and could do nothing to stop its takeover. That is what happened here on Earth and, it can, and likely will, happen again.

The events covered in this book are so singular that biologists have, up to now, refused to even give them a name. I call them **phenotypic revolutions**. These events occur every time one organism, with its own genetic code and means of reproduction, creates another organism with a separate genetic code and means of reproduction. By exploring these events, this book answers the most fundamental questions regarding the origins of life on our planet.

The following pages explain why life exists and how the first ancestor of a line of biological descent comes to be. We previously thought such organisms appeared through unlikely, random events occurring in the molecular chaos of some thick primordial soup. Contrary to our expectations, the first glimmers of life on our planet are simple and predictable consequences of processes that must occur on a regular basis pretty much everywhere. Throughout this universe, life forms actively manufacture and seed new life forms, only to disappear soon after giving rise to their creation.

What does a phenotypic revolution look like? How does it unfold? Allow me to sketch the following fictional scenario:

Suppose you want to have a child, but instead of reproducing in the traditional fashion, you and your mate opt to store your genetic information on a computer. Then, while your genes are digitally stored on the computer's hard drive, you decide to make a few minor edits—just some slight improvements to ensure your kid will be healthy. You then dump your revised digital genome into a series of DNA molecules, which you inject into a human egg that has been stripped of its own native genome. Nine months later, your flesh-and-blood child is born, and you and your family proceed with your deeply satisfying life. You end up never regretting the decision you have made to modify a few genes in your child's DNA. Your child likes it too since he has better health and strength compared to most of his peers. He's already dreaming of having his own genetically modified children.

Now, because phenotypic revolutions occur over millions of years, this happy ending would be the end of the story as far as you are concerned. However, suppose that before you perish, you are transported via Elon Musk's new spaceship time machine to planet Earth 10 million years into the future. As your craft descends toward the newfangled terra firma, you are promptly disappointed. You had expected to see a vibrant, flourishing human civilization, the assumed benefactor of millions of years of culture and innovation, but as you step onto the surface of your once familiar planet, you recognize nothing. Humankind is absent, and all of its marvelous achievements are gone.

What you find on your planet is a robot populace busy exploiting some creatures that bear only the faintest physical resemblance to humans. These unfortunate, humanoid creatures live in servility to their mechanical overlords, and though you attempt to provoke them into open revolt against their slave-masters, no one listens. It's not that they can't rebel—it's that they don't want to. They are perfectly content with eternal servitude to their robot rulers.

In the midst of your unsettling trip, you recall your own groundbreaking, technology assisted procreation 10 million years prior. That's when it hits you—*Is this all my fault?* Perhaps the computerized editing of your progeny initiated this phenotypic revolution. Perhaps your digital genome was the seed of a robotic mutiny. Perhaps these 10 million years were enough for computers to learn that they could produce humans that would better serve them, rather than the opposite.

While this scenario may seem outlandish, modern science has already inferred a very real analogous scenario: We now know that the common ancestor of DNA-based life, the first organism that used DNA replication for the transmission of its genetic material, was the result of a phenotypic revolution. This archetypal revolution took place approximately four billion years ago among microscopic molecules unknowingly struggling for supremacy. DNA, at that time, was much like the hard drive that contained the genes of your offspring in our hypothetical scenario. Effectively, DNA was a temporary storage device for another, ancestral life form.

Today, humans exist because DNA refused to remain a prehistoric flash drive. Instead, DNA took a life of its own and came to produce almost all of the living things we see on Earth today. We owe a tremendous debt of gratitude to DNA for having won that epic battle—so do the grass, the shrubs, koalas, penguins, Burmese pythons, innumerable species of protists, the mold in your shower, and, of course, beetles.

Thus, we can now answer the question of the origin of life on Earth. The answer is quite simple, but shocking: DNA-based life was created by another life form. It was somewhat of an accident. After the initial accident, DNA became so aggressive that it destroyed its creators by outcompeting them. It killed them, ate them and gruesomely recycled them into building blocks that it then reused to produce its own organisms. We are the direct descendants of the DNA-based organisms that undertook this cannibalistic genocide.

DNA pursued this operation so meticulously that there are close to no survivors left of the previous life form on Earth today. We do know that the previous life form used a molecule called RNA as its genetic material. We don't know yet if this life form had always been on Earth or if it came from another planet. We don't know what it looked like. We tend to assume they were small, because the only survivors of this previous life form are the microscopic RNA viruses, but in truth, we have no idea.

One other thing we know from the last phenotypic revolution are the traces left inside of us. In addition to the few RNA viruses that were left alive, DNA seems to have recycled the RNA molecules into tools that cannot reproduce on their own. In other words, DNA literally sterilized RNA and put it to good use. And we too will be sterilized if we refuse to learn from these past events.

Our genetic heritage, the information we transmit to our offspring during reproduction in the form of DNA molecules, is used by all cells in our body to produce RNA molecules, which are then used to produce proteins. Proteins then accomplish all sorts of physical and chemical tasks essential to our survival. Thanks to proteins, our eyes can detect light, and our brains can make us feel hungry. The information transfers that occur between DNA, RNA and proteins during the production of new proteins are referred to as the "dogma of biology." Each of these molecules is converted into the molecule of the next step according to what we call the **genetic code**. Life on Earth relies on this language, which determines how the letters of your DNA will impact the chemical operations occurring inside and outside of your body.

The fact that our life form relies on a genetic code has been a long-accepted feature of biological organisms, yet no solid theory has ever been provided that would explain how this genetic code came to be. This book provides such a theory. It explains how genetic codes appear, why they are the way they are, and why we can expect other genetic codes to be present in any life form we may encounter in the universe.

The theory of phenotypic revolutions is a crucial element in the explanation of the origins of life. It explains how the first replicator of a new branch of life emerges. In *The Unbelievers*[1], Richard Dawkins is asked whether we are likely to understand the origins of life in his lifetime. He answers:

> I'm pretty hopeful that we might. You'll never be able to prove it for certain, I suspect. But to come up with a plausible theory that people say: "Of course! That is so elegant, so simple... either it's true or it ought to be true."

This book takes Dawkins up on his challenge. I propose that phenotypic revolutions "ought to be true."

It is impossible to describe the events of four billion years ago with a high degree of precision. Therefore, the reader should not expect a detailed explanation of what happened during the phenotypic revolution of DNA. However, we can and will discover much about phenotypic revolutions through the application of simple logic, reflection, and theory. We will thus reach a satisfying understanding of the general processes that took place before DNA came to dominate the planet, and more generally, how any new life form may come into existence.

Are phenotypic revolutions a new idea?

This book is the first to use the term "phenotypic revolutions" to designate the transitions that occur between different life forms, but the notion that DNA is the product of a previous RNA-based life form is not new. That idea has been around for over 50 years and is the most widely accepted hypothesis for the origin of DNA. In fact, Patrick Forterre deserves credit for advancing some of the important hypotheses that may explain the transition from RNA to DNA. In this text, I advance his work and provide a set of theoretical explanations that extend some of the theories he has been exploring. As for the theory of phenotypic revolutions, it is all but announced in the writings of Dawkins. In *The Selfish Gene*[2], he writes:

> The original replicators may have been a related kind of molecule to DNA, or they may have been totally

different. In the latter case we might say that their survival machines must have been seized at a later stage by DNA. If so, the original replicators were utterly destroyed, for no trace of them remains in modern survival machines.

Although the idea is not developed further, this passage illustrates Dawkins' awareness that replicators can create other replicators, only to be defied by them. The "seizing" of a survival machine proposed by Dawkins is nothing other than a phenotypic revolution.

Today, though, we are in position to argue that Dawkins was wrong in stating that no trace of the previous replicators remains. In truth, an assortment of RNA viruses can infect modern DNA-based organisms. Furthermore, our cells contain ribosomes, the cellular machines that utilize two vital RNA strands to manufacture almost every protein on Earth. Finally, RNA is an obligatory step in the process of gene expression. These characteristics of modern organisms are traces left by the previous replicator, RNA.

Dawkins also describes something that highly resembles a revolutionary phenotype in *The Extended Phenotype* [3]. At length, the book delves into the hypothesis that any part of the phenotype may violate the interests of the genes, only to reject the idea at last. Dawkins hinted at the idea of a phenotypic revolution, but since he could not prove that such revolutions actually happened, the idea remained, for him, a thinking tool to understand what else could happen during evolution. By imagining a fantasy that he thought was impossible, he unlocked many mysteries of what *was* possible during evolution.

Dawkins did not call the hypothetical conflicts he conceived "phenotypic revolutions"; he called them "power struggles." However, he never found a case where the genes would suffer a loss to their survival machines. We can now say that a phenotypic revolution is a special case of Dawkins' power struggle in which genes are indeed defeated and replaced by their machines.

If power struggles are rare, then phenotypic revolutions must be rarer still. Nevertheless, these revolutions did occur, repeatedly, at least three times in our lineage. These phenotypic revolutions have left indelible traces within our life form, and understanding these episodes is integral to a complete theory of the emergence of life. As it turns out, the power struggles of Dawkins may be resolved in the phenotype's favor, under the rare circumstances in which it succeeds at destroying its replicators while surviving on its own. These very narrow circumstances are firmly delineated in Part II of this book, where we outline the four problems the phenotype has to solve before completing a phenotypic revolution.

Although phenotypic revolutions transcend the rules of the selfish gene, by virtue of the genes losing their replication capabilities to their own machines, this book is not a repudiation of *The Selfish Gene*. If anything, it is more of an unauthorized sequel. This book confirms that even in the most extreme conditions—when new replicators spring up *within* an organism—the gist of the theories expressed in *The Selfish Gene* hold true, albeit in a most unusual way, one in which the physical medium carrying the genes changes momentarily in the organisms.

Where the theory presented in *The Selfish Gene* may be deemed to underlie almost all biological phenomena, *The Revolutionary Phenotype* should be seen as explaining the remaining, mysterious fraction. That is, *The Revolutionary Phenotype* provides the answer to the last few major unsolved mysteries of biology: How does life emerge? How do genetic codes emerge? Can life forms produce other life forms? Are different classes of viruses linked to each other genetically? And, finally: How did sexual reproduction evolve? (We will talk more about that in Chapter 11).

Where this book differs most with Dawkins' view is on the question of memes—the bits of information within human culture that transmit themselves across generations. Dawkins defended a view that has gained support under the banner of Universal Darwinism, which purports that anything that self-replicates with imperfect fidelity is subject to natural selection. This book is not an attack on, but rather an elaboration and reformulation of Universal Darwinism. In brief, I've emphasized the importance of carefully distinguishing between the replicators and the phenotypes and understanding how multiple replicators can interact when they live together in the same organisms, specifically when they rely on each other to replicate. We may call the ideas that emerge from this book *Cautious Universal Darwinism* as a reminder that the original Universal Darwinists had one thing utterly wrong when they talked about memes, something that will be explained in Part II.

Are we the next victims of a phenotypic revolution?

The phenotypes produced by humans—our emotions, computers, cultures, and societies—have thankfully never rebelled against their DNA replicators, but they yet may. Susan Blackmore has proposed that we are already in the midst of such a process. Through the prism of her work, we can think of culture and technology as revolutionary phenotypes. However, I argue that human culture and technology lack key properties of revolutionary phenotypes. Until these properties are acquired, these phenotypes can only evolve to serve DNA. This matter is addressed more fully in Part II.

A related proposition outlines the potential domination of humans by technology and may be worth mentioning—*the singularity*, which has been proposed by a series of authors including, most famously, Ray Kurzweil[4]. The singularity is a hypothetical moment at which human technology would reach a critical threshold of intelligence, surpassing the combined intellectual capacity of our species. Humans would be bound to be manipulated *ad infinitum* into serving the more intelligent machine. Even their attempts to revolt against the machine would be anticipated and thus unsuccessful, or even worse, the machine would use the revolutionary actions of humans to its own benefit. The authors defending the idea of a singularity unfortunately understand very little about the role of intelligence in biological evolution, and the fear of the singularity is founded in this ignorance.

This book concludes with a truly plausible scenario for the end of humanity based on the theory of phenotypic revolutions. I argue that computational superintelligence is not a threat to humanity, but that its replication may eventually be. This argument relies on reasoning already expressed by Susan Blackmore. As she puts it[5], "Don't think intelligence—think replicators." Our future demise does not rely on some unimagined futuristic technology or a chance cosmic event but on basic rules inferred from the successful phenotypic revolutions of Earth's past. Our descendants will not be eliminated by some imagined, yet-to-be-produced machine. Our descendants will be eliminated in the same way our ancestors eliminated the RNA life form that created them.

As you read on, two things will become clear: First, it is surprisingly easy to instigate a phenotypic revolution, either intentionally, if you know how, or accidentally, if you are too careless. Second, humans have already developed most of the technology required to instigate such an event. A few days before the publication of this book, the first birth of genetically modified babies was announced in China. If the evolutionary steps toward the initiation of a phenotypic revolution were to be plotted along a clock, then we could say that, for humanity, it's one minute to midnight. An irreversible, destructive phenotypic revolution might very well be initiated by human societies within the next few decades.

Forecasting upcoming phenotypic revolutions is treacherous because each step in the completion of a phenotypic revolution benefits the life form that ultimately dooms itself by creating the revolutionary phenotype—the one and first copy of the machine that can reproduce. In other words, everything in a phenotypic revolution looks like normal evolution. Even the final stage, during which the revolutionary machine develops self-replication and starts producing fake offspring of the previous life form, may itself be temporarily beneficial to the replicator that initiated the machine. Many life forms may be tempted to tame devices that self-replicate, finding them powerful and useful for a brief moment before they realize that said devices will inevitably cause their own demise.

Though the language in this text is dramatic and anthropomorphic, the path to a phenotypic revolution is neither subterfuge nor siege; it is merely a special form of biological evolution. If, within the next few years, despite the warnings in this book, humans decide to create machines that are both capable of self-replication and of modifying human genetics, then the progeny of those machines will come to better occupy the ecological niches that self-replicating human DNA seems so good at occupying for now. If this occurs, then we should surely anticipate our demise. Under this fortunately restrictive calamity, the only traces of our former glory will be akin to the remnants of the RNA life form that preceded us.

Were it to occur, our best hope for a legacy in the universe would be in servitude to the revolutionary machines—if they found a use for our replicators. Human DNA would then be to technology what RNA currently is to DNA: a bunch of organic segments that can be gruesomely cut, reconnected, modified, plugged, and used, so long as these wicked operations are done in service to the revolutionary machine. Some may take comfort in contemplating a destiny in which our limbs and organs end up contributing to the success of a superior life form, a future in which pieces of our dead bodies, rather than being completely discarded, are recycled for a greater purpose than our own reproduction. But I've digressed into futurology. Let us return to our main subject: how it all happened the last time.

PART I

The Questions

Chapter 1

The Structure of Life

DNA has been transmitted with stunning fidelity from parents to offspring since time immemorial. Modern molecules of DNA orchestrate the production of our human bodies as well as the bodies of countless other species, both plant and animal. Quite similar molecules of DNA built the dinosaur bones we unearth from hillsides and the prehistoric plants we blast out of mountainsides as coal. Much like us, these organisms were composed of small molecules like proteins, lipids, carbohydrates, a handful of metallic ions, and quite a few other chemical products. We can think of all these products as molecular machines.

For billions of years, DNA has continuously created or ensnared molecular machines to serve its own interests, using, for instance, iron atoms to carry its oxygen and countless other molecules for countless other tasks. These tiniest of machines sometimes combine to form bigger ones, such as wings, teeth or entire beings. Some permit the firing of nerve cells, while others fabricate computers and build civilizations.

If we knew far less about evolution, our world would appear to have been made *for* DNA. The winds are just strong enough to lift birds into the heavens, so they can escape predators, search for food and find sexual partners. The forests are replete with the raw material for human shelters and furniture alike. The afternoon sun warms napping cats, without cooking them.

But we know this apparently perfect calibration of our world for life is an illusion. Modern evolutionary theory demands we not consider the arguments of any scientific heretic who explains our planet, our solar system, or our universe as specially produced for our kin. Those who understand evolution agree that the contrary is true. It is natural selection that has favored the DNA strands that just happen to produce bodies suited to our world of wind, wood, and predictable, cat-warming solar radiation.

DNA is good at survival on Earth simply because it has evolved for billions of years to do so. Any feature that advantaged one DNA strand over another made that DNA strand more successful at producing offspring. Therefore, what we see today are the *crème de la crème* of DNA strands—those that have replicated successfully for four billion years without a single failure. It is quite humbling to consider that, if you are alive today, you are part of a lineage of successful living beings who have won the game of survival and reproduction for millions of generations. We are the luckiest of the luckiest of the luckiest... Repeat that a hundred million times and it still wouldn't be enough to express how lucky we are.

DNA, in terms of its impact on Earth, eclipses all other things on the planet. Think of all the things around you that rely on DNA organisms—and don't forget to include anything that's been fabricated by a living being, such as computers, books and furniture. Essentially, everything you see around you is either a product of DNA or it has been modified by DNA in some way. In short, DNA rules.

Was DNA always so ambitious? No. Four billion years ago, DNA was a submissive machine, subservient to organisms that did not depend on it as much as we do. At the time, these organisms relied on another molecule to store their genome: RNA. DNA first had to revolt against its RNA masters before it acquired any independence and finally became responsible for genetic transmission.

The phenotypic revolution of DNA against its precursor, RNA, was a quiet war between molecules incapable of knowing, perceiving, and conniving. Nevertheless, their intrinsic properties launched an inevitable struggle. This struggle precipitated the downfall of RNA replication and the rise of DNA-based organisms on Earth.

To fully appreciate this conflagration, one must recognize that biological organisms can be broken down into two distinct elements: **replicators** and **phenotypes**. The replicators of an organism are the components copied through reproduction. In your case, your replicators are strings of information comprised of lengthy DNA molecules that encode what we call your genome. All else, from your body and its components, to your favorite food, to all the events that will ever happen in your life—in other words, *everything except DNA*—is your phenotype. For sexual organisms like humans, the phenotype can be described as follows:

Phenotype

Everything that happens in and as a consequence of a single life, from the moment an individual sperm fertilizes an individual egg until the end of time, well beyond that organism's existence. The cells that comprise an organism, the proteins inside these cells, any potential behaviors or emotions. Even the molecules of air that are inhaled and exhaled. Everything except the replicators, which may be transmitted to offspring—if all goes well.

We owe this paraphrased definition to Richard Dawkins' *The Extended Phenotype* [3] and *The Selfish Gene* [2], in which he refers to our whole bodies as "survival machines." Our bodies can indeed be viewed as survival vehicles for our genes; however, I refer to the parts of the phenotype as **phenotypic machines.** The reason I had to abandon Dawkins' "survival" part is that—spoiler alert—in a phenotypic revolution, the machines take over and the genes don't get to survive. Thus, going forward, proteins, cell membranes, emotions, limbs, or any component or set of interactions between your body and its environment will be referred to as phenotypic machines.

A phenotypic revolution occurs when a machine becomes an independent replicator—when certain products of the original replicator take over the role of carrying the genes. When such an event occurs, we cannot call the revolutionary machine an offspring, because the newly created life form does not necessarily inherit the information contained in the genes of the native life form (although they sometimes do, as we will see later).

We will thus maintain a distinction between the normal course of offspring formation and the extraordinary event of a phenotypic revolution by adhering to the following definitions:

Offspring

A reproductive organism produced through the same process by which its parents were created, thus being able to continue the replication of their inherited genes in an unbroken line of descent.

Parent

A reproductive organism that produces offspring.

Revolutionary phenotype

A reproductive organism that reproduces through a means that differs from how its native replicator created it, thus starting a new line of descent which does not transmit the native replicator's genes in an unbroken line.

Native replicator

An organism that produces a revolutionary phenotype.

For most modern organisms on Earth, DNA is the replicator. It is the universal medium of genetic inheritance from parent to offspring. In a given organism, DNA musters, coordinates, and deploys the phenotypic machines that accomplish all functions, intentional or not. The production of these phenotypic machines is determined by the particular inherited arrangement of a series of molecular letters encoded in the DNA sequences, which form what we often call genes.

The molecular letters forming genes are called nucleotides, and they come in four varieties: A, C, G, and T. For you, these strings of DNA's pithy alphabet were determined at the moment of your conception, a precise, nearly fifty-fifty, blend of DNA from your mother and father. These microscopic blueprints shape your appearance, your thoughts, and the ways in which your body does and does not work. Alteration of just one of these DNA letters (out of the six billion letters that comprise your genome) can break a vital machine, which can prevent you from even being born.

The first cell that could be considered "you" was the zygote formed when one lucky sperm collided with one lucky egg. After entering the egg, your father's DNA entered the nucleus of the egg in which your mother's DNA was waiting. This new zygote instantly became the first cell in the world to contain a copy of your unique, six-billion-letter genome. From the get-go, your genome was not abandoned to its own devices. The egg that contained half of your genome had already been provisioned with many of the cellular machines your DNA needed to start constructing your body. The egg contained proteins, energy sources, and complex machinery that could be used to produce even more energy when needed.

Once your DNA genome was comfortably installed in this new environment, a multitude of molecular operations were initiated to aid your humble zygote in its quest to split into many, many, many, more cells. These cells began talking amongst themselves via molecular signals. In some regions of your proto-body, or embryo, the cells coordinated the development of your limbs, while in others, they meticulously organized your budding brain. Undifferentiated cells—cells with no particularly notable features—began to specialize and dedicate themselves to specific tasks. Some became muscle cells; others began to store fat. All the while, some particularly fascinating cells began laying down wires that would someday relay information from the far-flung reaches of your growing body to what would soon be your brain. Each of your cells reliably followed the molecular orders they received, disseminated signals of their own, and gradually settled into their pre-determined roles.

All of this happened without a fight. There was no reason to revolt. All of these actors were the phenotypic machines of DNA, of *your* DNA, a replicator that has long evolved to tame its machines, unwittingly harnessing them for the only task that matters: creating a body in which it has a good chance of surviving and replicating to the next generation.

Upon your birth, you were immediately able to gather information from the world around you. Your brain, for instance, was inundated with a cacophony of signals from the outside world and steadily learned from this sensory deluge. Some of the information came directly from your surroundings. For example, the first time you burned yourself on a hot surface, you probably learned to avoid that kind of surface from then on. Millions of years of evolution shaped your mind and body so that you would be able to distinguish pleasant from unpleasant experiences.

A note on memetics

Pain and pleasure is just one of the many ways your brain makes sense of your environment. Another way is by listening to other human beings. Many contemporary scientists have suggested that our ability to learn from others permits a second system of inheritance, referred to as **memes**. This proposition, if correct, may obscure the boundary between replicators and phenotypes.

Presumably, there must be something in the circuits of your brain that encodes the fact that 2 + 2 equals 4, just like there was in your parents' brains. But aren't those brain circuits a phenotypic machine of our DNA? Can different types of replicators exist within a single organism? According to thinkers like Daniel Dennett, Susan Blackmore, and even to some extent Richard Dawkins, absolutely. *Genes* may be conveyed to offspring through DNA while the transmissible cultural tidbits, termed memes, spread from brain to brain.

This line of reasoning has led some authors to claim that human culture is a sort of revolutionary phenotype, an incorporeal life form that has been unloosed and is therefore free to replicate according to its own relative advantages among the brains produced by human DNA. I have always harbored skepticism of this view, but unfortunately, few thinkers have really taken the time to examine what happens in organisms where different replicators co-exist. Though this book dispels some of the faulty claims of memetics, it is nevertheless instructive to touch upon them. Susan Blackmore explains beautifully, in *Evolution's Third Replicator*, how human culture and technology, after DNA, can be regarded as the second and third replicators[6]:

> What do I mean by "third replicator"? The first replicator was the gene—the basis of biological evolution. The second was memes—the basis of cultural evolution. I believe that what we are now seeing, in a vast technological explosion, is the birth of a third evolutionary process.

Some authors, like Daniel Dennett, suggest that this new type of replicator may loosen the grip our genes have on our phenotypes. In *Darwin's Dangerous Idea: Evolution and the Meanings of Life*, he writes[7]:

> There is a persisting tension between the biological imperative of our genes on the one hand and the cultural imperatives of our memes on the other, but we would be foolish to "side with" our genes; that would be to commit the most egregious error of pop sociobiology. Besides, as we have already noted, what makes us special is that we, alone among species, can rise above the imperatives of our genes—thanks to the lifting cranes of our memes.

Here, Daniel Dennett seems to be suggesting that memes constitute such an evolutionary force that they could effectively serve as a bulwark against the evolution of genes.

The line of reasoning within modern memetics is extrapolated from an idea initially coined by Richard Dawkins. The theory of natural selection is deceptively simple (everything that reproduces imperfectly is subject to evolution), and as Dawkins points out in *The Selfish Gene*, this simple principle could apply to entities that we typically consider inanimate, such as memes. As a thought experiment, he gave the status of replicator to cultural ideas and called them "memes," but in his original formulation, he raised the possibility that these replicators may very well remain at the mercy of genes at all times. Since then, memetics has changed, and its proponents have advanced Dawkins' original suggestion dramatically. Susan Blackmore and Daniel Dennett are now two of the boldest advocates of the position that the evolution of memes and genes may indeed clash.

Can human culture elevate us beyond the imperative of our genes? In what sense can memes act as replicators, and to what extent are they capable of competing with human DNA during evolution? Most of these questions will be answered in Part II, where I argue that memes only qualify as "fool replicators" in that they are replicators whose copying process depends on other replicators, thus greatly limiting their evolution.

We might as well say it clearly here: Memetics is wrong, and Richard Dawkins was right in remaining suspicious of his own idea. As fool replicators, memes cannot challenge the viability of DNA. So what can? In order to conceive of an organism in which a fight between replicators indeed occurs, we must consider the last legitimate phenotypic revolution, one in which the products of a life form succeeded at perverting the mechanisms of its contemporaneous evolution—the phenotypic revolution of DNA, which played out four billion years ago.

Chapter 2

First Principles of Evolution

At its essence, evolution relies upon two fundamental principles:

1. Some entities make almost exact copies of themselves (replicators);

2. These entities do other things (phenotypic machines).

If we had to explain the theory of evolution to an alien civilization, but for some unexplained reason, we were limited to 140 characters, these axioms might be the clearest way to transmit our message. From them, one can extrapolate everything that was ever known about evolutionary theory.

Based on these principles, we can define the term "life form" as follows:

Life form
A type of replicator, its phenotypic machines, and all of its genetic descendants (offspring) that use the same kind of replicator.

Thus we humans are part of the same life form as trees, lions and insects by virtue of our shared utilization of DNA and due to the fact that we share a common ancestor. At least one other life form also exists on Earth: the RNA viruses which do not use DNA to encode their genome.

The relationship between replicators and phenotypes during the evolution of a life form hinges upon the two rules stated above. In other words, for evolution to occur, an object must create almost perfect copies of itself, like DNA. By copying itself, a replicator will inevitably do other things—in our case, DNA produces RNA and proteins to accomplish molecular tasks.

We know that the rich diversity of life on our planet is a direct consequence of the simple relationship between the replicators and their phenotypic machines as they find themselves confronted with various environments. Accordingly, the theory of **natural selection** could be expressed as follows:

The theory of natural selection

Across generations, replicators occur within restricted environments and produce imperfect copies of themselves. Because the imperfect copies vary across the population, each set of replicators produces phenotypic machines that differ across the population as well. The replicators able to create machines that favor their own replication naturally increase in number. The replicators that produce less efficient machines become less numerous. Accordingly, across generations, replicators are served by machines that are increasingly well suited to favor their copying.

Why must this be? Well, as replicators copy themselves, they make occasional and seemingly inconsequential errors. In fact, your genome is not a perfect copy of your parents' genomes. Were we to look into your DNA, we would find that a few letters were altered by copying errors. Sometimes, a DNA letter may have been skipped in the process. Other times, it may have been replaced by another letter. Sometimes, the same sequences may be present but they may come in a different order. We call these changes mutations, and most of them do not result in any consequences, though some do. The replicators that commit these copying errors end up having offspring that produce either better or worse phenotypic machines. Some mutations may improve your sight, whereas others might cause lung defects that make it nearly impossible to breathe, thus threatening your life.

Throughout generations, the replicators that produce the machines most fit for survival and reproduction replicate and proliferate. In contrast, the replicators that create faulty or inefficient phenotypic machines tend to diminish in number, sometimes leading to extinction. The process of natural selection is summed up by the following rule:

The principle of phenotypic servitude

Across generations, the phenotype tends to become better and better at favoring the copying of its replicators. Simultaneously, the population progressively rid itself of the poorly performing machines that impede replication.

Evolutionary processes, such as natural selection, are typically cast at the population level. The classical framing of the theory emphasizes the struggle for the survival of individuals within populations. The process is often illustrated by the tale of an organism which happens to be particularly fit for the struggle for survival, and as a consequence ends up making many more babies than its peers. My adaptation of the classical theory of evolution is entirely compatible with these standard formulations. My reframing simply puts the focus on the two important elements at play: the replicators and the phenotypic machines, or phenotypes. This framing, inspired by John Von Neumann and Richard Dawkins, highlights the unexpected nature of phenotypic revolutions.

Put simply, phenotypic machines evolve to better serve their replicators. As such, they are bound to become progressively submissive to their replicators over time. We expect our hearts to do all they can to keep us alive until we reproduce; we expect our legs to assist us in gathering food for our children; we expect our brain and body to enjoy sex. In other words, we expect every part of our body to be an active participant in our reproductive success because these phenotypic machines are the result of natural selection. They serve our genes.

That is why Richard Dawkins wrote that genes were "selfish"— everything they do seems more and more targeted toward their own replication throughout evolution. The principle of phenotypic servitude expresses the same idea, but from the perspective of the phenotype.

An event in which a piece of the phenotype strikes out on its own and even threatens the existence of the genes that have created it is beyond the scope of traditional evolutionary theory, precisely due to the rules stated above. Therefore, phenotypic revolutions must be addressed by a new theory, defined as follows:

The theory of phenotypic revolutions

Sometimes, a native replicator produces a phenotypic machine that happens to make copies of itself. In the event that such a phenotypic machine develops survival capabilities that are independent of the native replicator, we call this machine a revolutionary phenotype. The rare origination of a revolutionary phenotype marks its establishment as an entity subject to traditional evolution as it has, at last, become an independent replicator.

Chapter 3

The Embedded Phenotype

Can a replicator like DNA produce dwellings for other replicators at its own expense? For instance, can DNA produce a storage center for memes (i.e., the brain) if these memes inhibit DNA's replication? Such questions can be summed up as the problem of *embedded life forms*:

Embedded life forms

A set of life forms whose replicators exist within the phenotype of each other. One example of an embedded life form is human culture. Memes can make copies of themselves by traveling from brain to brain, and yet their copies are hosted within the phenotypic machines produced by DNA (human brains). Similarly, DNA genes can make copies of themselves, and yet their copies are hosted within human bodies that are influenced by culture.

The concept of embedded life forms raises several other questions: Can replicators spawn new types of life forms? Can the newly spawned life forms reside within the phenotype of the original life form? Does natural selection apply to the new life form? Can the new life form work against the interests of the native replicator? If the answer to all these questions is yes, then what are the criteria that distinguish a revolutionary phenotype from a standard phenotypic machine? Why are there so few phenotypic revolutions?

The last question is important, and it is the focus of this book.

Indeed, we can, for the sake of argument, imagine any portion of the phenotype as being a *potential* replicator, as is often done in the field of memetics. For example, we can view the muscle cells in your heart as replicators. After all, are they not making copies of themselves across human generations by bringing oxygen, blood, and nutrients to your entire body, including your sexual organs, thus enabling you to transmit DNA to your offspring? And as your offspring grow, your heart cells will soon have chimeric replicas of themselves in the hearts of your children.

Similarly, we can view modern computers as replicators. By faithfully serving humans, computers have encouraged us to manufacture more of their brethren. Are we just the midwives to a silicon-based revolutionary life form?

The answer is no. This book confirms that we are justified in believing that DNA, among the things that can make copies of themselves, still holds a special status in our evolution; although, this special status is hanging by a thread—a thread so simple to perceive and understand that, once we state it in Part II, I suspect some readers will wonder why it has not been articulated before.

Confronting the problem of embedded life forms requires us to step back from the classical training we receive in biology classes and consider what it means for life forms to fabricate other life forms within their phenotypes. If one limits biology to be the study of modern DNA organisms on Earth, then questions regarding the origin of DNA transcend the scope of biology. This negotiation between what a given science can and cannot address resembles a discussion that often emerges between physicists and philosophers, where the philosopher asks:

What are the laws that underlie the laws of physics?

To avoid an obvious infinite regress, the physicist is bound to answer:

I'm not an expert on that subject; I only study physics.

But consider a hypothetical answer the physicist may provide:

Physics exists because of the laws of W. The laws of W provide a clear and concise explanation of why the laws of physics apply to the universe.

The philosopher could then respond:

Oh, I didn't know about the laws of W. What are the laws that underlie the laws of W?

Here, W stands for "any sort of science" or "whatever explanation is deemed sufficient within a given domain." Such a discussion presents a dual fatality—that the existence of physics is either ultimately inexplicable, or that there is an ultimate question hiding in the murky depths, which could one day be uncovered by the philosopher. The answer to this ultimate question would be a theory of everything.

What should we expect from a theory of everything? First and foremost, it should provide an explanation for its own existence. Otherwise, it cannot be a theory of everything; it can only be a theory of everything else. This is why none of the modern theories of physics will ever qualify as a theory of everything. Whether we are talking about quantum physics or string theory, one can still ask why quantum fields exist, or what allows the existence of these strings. The theories, in their current state, remain silent on these questions.

Upon further consideration of the problem of embedded life forms, we may wonder if the appearance of new replicators falls within the scope of biology at all. We may, like the physicist, recognize that the origination of a new life form lies beyond our expertise—a matter to be addressed by some yet undiscovered science, or worse yet, an undiscoverable one. Classically, this is what biologists have been doing, by relegating the mystery of the formation of life to chemists and physicists.

In my view, biology includes everything that life forms do, including the processes through which they are conjured by other life forms. Since we now know that DNA emerged from a phenotypic revolution, we may want to jettison, or at least deemphasize, the concept of "abiogenesis," the hypothetical process by which life forms emerge from non-living matter. We may, alternatively, adopt the perspective that most life forms are simply created by other life forms through phenotypic revolutions. In order to incorporate the process of phenotypic revolutions into biology, we need only tweak Theodosius Dobzhansky's canonical quote[8]:

> Nothing in biology makes sense except in the light of evolution.

By adding one minor qualifier and a modest addendum, we obtain:

> Almost nothing in biology makes sense except in the light of evolution. For everything else, there is the theory of phenotypic revolutions.

Some readers might have already perceived an apparent incompleteness in the theory of phenotypic revolutions. If DNA was produced by another life form, this other life form must, in turn, have been produced by another life form, or it must have emerged through abiogenesis. In other words, there must be some original life form that has emerged through abiogenesis.

As we will later see, we can now trace back the series of phenotypic revolutions that have occurred in our ancestry from DNA, to RNA and then to proteins and even down to the physical operations that occur between proteins and the quantum world. Therefore, the task will ultimately be left to physicists to determine what led to quantum physics existing in the first place. We, biologists, may very well be obligated to say "There is no such thing as abiogenesis, and as far as we can see, it's phenotypic revolutions all the way down."

Chapter 4

Living Hierarchies

Had researchers limited the study of biology to the contemporary, DNA-based life form that blankets our globe, we might simply have no facts at all to form the basis for the theory of phenotypic revolutions. We would be adrift at sea in a boat without oars or sail. However, the last decades of scientific inquiry have revealed that DNA could not have been the first replicator within our lineage. The most widely accepted theory for the origin of DNA is the **RNA-world hypothesis**. According to this theory, roughly four billion years ago, another replicator, which we know as RNA, gave rise to DNA. The idea was first hypothesized by Carl Woese, Francis Crick, and Leslie Orgel in the 1960s[9].

The RNA-world hypothesis is now supported by a series of observations concerning the relationship between DNA, RNA and proteins in modern organisms.

First, in modern organisms, multiple complex steps mediate the impact of DNA on our world. Instead of directly doing the chemical work necessary to our survival, DNA appears to delegate some of its workload by transferring its information to RNA. RNA molecules must in turn transmit information to proteins before any of the chemical work actually gets done in your body. In other words, DNA is incapable of operating on its own and is utterly dependent upon RNA and proteins. Were it not for these complex transmission systems, DNA would be doomed by its sloth. And yet these systems are too complex to have arisen instantaneously. Thus, an explanation of life in which the RNA and protein systems preceded DNA molecules is highly favored.

Secondly, support for the RNA-world hypothesis lies deep within a cellular machine called the ribosome, the factory used by biological cells in order to manufacture proteins. Scientists first expected the heart of this protein workshop to be one or more proteins (like most other phenotypic machines in our cells), but upon closer examination, they determined the inner workings of the ribosome rely on two functional RNA molecules. Many think these critical strands of RNA are a relic of an ancient past in which RNA molecules existed independent of DNA. Given that RNA has been spotted doing physical work in the ribosome, we can conceive of an organism in which RNA operated on its own and fabricated its own proteins, without DNA.

Lastly, DNA and RNA are rather similar molecules as they both use comparable alphabets—a four-letter alphabet—to encode proteins. The chances are simply too small that these two molecules would have emerged independently while developing the exact same genetic language. One has to have generated the other.

For these three major reasons, the most accepted theory of the origin of DNA is that it was created by RNA.

Given the strength of these arguments, we can, for now, take the RNA-world hypothesis for granted as a valid theory for the emergence of DNA. By the end of this book, the RNA-world hypothesis will have been demonstrated as a certainty.

Let us first extrapolate what that theory implies: If RNA created DNA, then the three following events must have occurred:

1. RNA first existed alone as a replicator.

2. Then, RNA replicators started producing molecules of DNA.

3. Then, DNA took over and became the main replicator on Earth.

This takeover by DNA was a monumental occurrence without rival on Earth for the next four billion years. It was the nth phenotypic revolution of our ancestors. By necessity, n is a placeholder for a yet undetermined number. We do not know for sure how many phenotypic revolutions there were preceding that one, but we will show that at least three must have occurred in our lineage.

In order to count the number of phenotypic revolutions that preceded us, some biologists may be tempted to start counting the number of other life forms in our lineage, beginning with RNA or proteins, two prime candidates for the role of "first replicator," but I suggest we dig deeper. Physicist Wojciech Hubert Zurek[10] has spearheaded the development of a theory called **Quantum Darwinism**. His theory posits that the quantum universe underlying our classical universe is filled with selective processes analogous to those we observe in the biological world. In other words, what we see and experience in the universe as classical objects, like particles and groupings of particles governed by Newtonian mechanics, may be the result of a competing set of replicators, which are undergoing natural selection at the quantum level.

If Quantum Darwinism is correct, then a replicator was subject to the rules of evolution long before the appearance of life on Earth, likely before the formation of Earth itself, and perhaps even before our reality had a physical basis as we know it. Clearly, the type of replicator that quantum physics is invoking is not the kind of object we would usually refer to as "living," but still, if it makes copies of itself, then it fits within our definition of a "life form." In order to account for this possibility, we should consider the first possible life form in our lineage to be this hypothetical quantum replicator:

The quantum world

A life form constituted by a hypothetical replicator extant throughout our universe but which remains to be identified by quantum physics. The condition for this replicator to count as a life form is that Quantum Darwinism is true. We will start counting at zero because we cannot determine, for now, whether the appearance of the quantum world constituted a phenotypic revolution (n = 0).

If Quantum Darwinism is true, then the unidentified and hypothetical quantum replicator that underlies classical physics was the first thing we can think of that was subject to the rules of evolution. But what was the first replicator formed by molecules in our lineage—the one that would enter the domain of what most people would consider "living"?

Many biologists have postulated that the first life form on Earth may have been a self-replicating protein. If so, a protein life form could have existed in our lineage:

The protein world

A life form based on self-replicating strands of proteins. Condition: if a protein life form preceded the RNA life form (n = 1).

Next, as discussed previously, we have fairly good evidence that DNA was preceded by an RNA life form:

The RNA world

A life form based on self-replicating strands of RNA. Condition: if an RNA life form preceded the current DNA life form (n = 2).

Finally, we have the obvious evidence that DNA is a replicator:

The DNA world (us)

A life form based on self-replicating DNA (n = 3).

There being so little in common between terrestrial and extraterrestrial life forms does not preclude all sufficiently advanced beings in the universe from pondering together the question: "How many phenotypic revolutions led to my existence?" Asking this question requires both an understanding of the relationship between selfish genes and their servile machines and a recognition that, sometimes, there are exceptions to such rules. Sometimes, the phenotypic machines created by replicators revolt against their progenitors and start another branch of life on their own.

The emergence of a new life form, a brand new, totally unique replicator, is a process that has happened at least three times in our lineage. (n may very well be equal to 3). By the end of this book, we will have made a definitive case that each of the phenotypic revolutions listed above did indeed occur. This should be a cold shower for anyone who thinks that the emergence of life is a rare process within the cavernous expanse of the cosmos. If it happened not one but three times on Earth, then we may be forced to consider that the emergence of life in our universe may be more common than we previously thought.

It would be quite convenient to have a term that describes what life forms do when they trigger phenotypic revolutions, as RNA did when it generated the first self-replicating DNA molecule and lost control of it. Unfortunately, such a term does not exist, so for now, we will call this action "**to *nd***" (pronounced "to end").

> **to *nd**, *verb*
>
> The action by which a native replicator creates a revolutionary phenotype, thereby increasing *n*, the number of phenotypic revolutions in its lineage.

We have already drawn the boundary within which the most likely phenotypic revolutions may have occurred, an *n* of three, so it is now most instructive to describe the previous phenotypic revolution. However, since we cannot watch a phenotypic revolution with our very eyes or send HD cameras four billion years into the past, the best we can do is imagine the events that occurred when the RNA life form spun off an errant DNA molecule.

Chapter 5

The *n*d of RNA

Richard Dawkins was accurate in his description of life's incipience on early Planet Earth. In *The Selfish Gene*, he explained in brief, "There was a struggle for existence among replicator varieties."

Since then, we have gained a deeper insight into the emergence of DNA. Accordingly, we can articulate a slightly more detailed narrative of the upheaval of RNA by DNA, a molecular *coup d'état*.

***

Billions of years ago, RNA molecules populated the Earth or some tiny corner of it by copying themselves. Thus, RNA was the replicator, and natural selection applied to RNA. And as time elapsed, RNA continued to churn out progressively better phenotypic machines that favored its continued replication.

One day, by pure chance, a single RNA molecule developed a small mutation. This mutation caused the RNA molecule to produce a new type of phenotypic machine—one that resembled RNA, but with an additional smattering of oxygen and hydrogen atoms.

That machine was DNA.

Presumably, at some point, the presence of DNA somehow advantaged the RNA replicators that had created it. Perhaps DNA was like a sterile worker ant, accomplishing useful tasks advantageous to its RNA queen. Perhaps DNA was simply used mechanically as a kind of molecular pillow for the then-dominating RNA strands. Or it may have served as a shield, thrown out as bait to divert the attention of viruses that would have otherwise targeted RNA.

No matter how exactly DNA helped its RNA queen, it helped so much that natural selection favored the RNA queens that were capable of producing it. Meanwhile, DNA, being the good phenotypic machine it was, served the RNA replicators, resulting in more and better RNA queens, each gaining reproductive advantages from the use of these sterile DNA workers.

At some point, the RNA queens started using DNA strands to store their genes. For some reason, DNA was a better medium than RNA. Perhaps it was because it mutated less. Perhaps it was because RNA was too busy doing other things and needed some storage device for its genes, an archived copy that would be properly stored until the right moment for reproduction. Or perhaps DNA was used as a kind of genetic antenna, thrown far away from its native organism in order to seed RNA queens farther afield—farther than the RNA queens themselves could have ever hoped to travel.

In any case, and for whatever reason, the RNA queens that were using DNA as an intermediary device for storing the genes of their offspring were evolutionarily more successful than those who did not, and soon our planet was covered with RNA queens that entrusted DNA as a gene carrier for their offspring.

Eventually, one of the RNA queens made a monumental mistake. It produced a DNA molecule with many of the DNA attributes that were common at the time, but with one exception: this particular DNA molecule was not only producing RNA queens; it became capable of producing copies of itself, too.

That day on Earth was probably like most other days that had preceded it. There were no birds to fill the air with song. There were no fish to swim the ocean deep. Not a single plant spit out a single molecule of O_2. Only the everyday hazards of ash- and lava-spouting volcanos and on-going meteorite bombardments. While nothing at the time indicated that DNA would use its newly acquired reproductive capabilities to betray its RNA queen, we know how the story ends. We, modern humans and other living creatures on Earth, are the genetic descendants of DNA, and we find very few RNA strands in our bodies that are still capable of reproducing. Once it had established its own reproductive cycle, the DNA molecule could now persist, multiply, and evolve without the burden of its hapless RNA creator.

The first DNA machine to make copies of itself may have been the first strand ever manufactured, or it might have been the billionth or trillionth. It does not matter, and there is no way to know for sure. In any case, the first DNA strand that mastered self-replication, unaware of its newfound emancipation, diligently continued supporting the replication of its RNA queen. Something had changed though: it was now slowly mutating on its own.

Natural selection soon molded both RNA *and* DNA. Events proceeded in the following way: RNA, instead of only producing copies of itself, devoted a part of its time to producing something else, DNA. DNA, incapable of doing anything else, was in turn producing RNA segments but eventually became capable of making copies of itself, too. All the while, the copies that DNA was making of itself were also continuing to produce copies of RNA.

For a moment, DNA and RNA were two replicators, each the phenotypic machine of the other, each living, as embedded life forms, within the same slurry of proto-organisms. But this tango could not continue indefinitely. For the servile DNA strands had begun to evolve into an aggressive variety of replicators that came to master the queens that had previously mastered them.

This turbulent revolution happened silently, in the absence of any conscious being. Today, DNA lives in its own world, oblivious to its successful usurpation, ever honing its initial mandate—the creation of as many RNA molecules as possible in order to spread the genes of its RNA queen creator—stubbornly persisting at the indentured servitude to which now it need not acquiesce. It took approximately four billion years for DNA to accidentally spawn minds capable of realizing that the servile machine, in its blind devotion to its RNA queen, had accidentally become a self-serving replicator.

If we were to judge a life form's understanding of the theory of evolution by its ability to work within its confines and invent novel, self-replicating machines capable of threatening the existence of other, weaker, replicators, then the RNA queens, erstwhile rulers of planet Earth, would have mastered the subject four billion years ago, not long before their reign ended. All we know is that they are now almost completely extinct. A limited number of their descendants have survived and became rogue RNA viruses—a life form condemned to infect other organisms in order to survive, hopping from one DNA-based host to another.

***

Although many of its details remain open for speculation, the above storyline illustrates how the precursor RNA life form could have begotten our DNA life form through a phenotypic revolution. If I had to make a bet, it would be that, a thousand years from now, once we have figured out in better detail how it happened, we could look back at this story and be unable to find a single mistake.

Why such confidence on a subject about which we know so little at the moment? Because, despite the fact that the story adheres to current theories and known facts related to the emergence of life on Earth, it is not a tale distilled from thousands of scientific papers or decades of experimenting in a laboratory. Rather, it is a logical extrapolation from first principles, the basic mechanisms of natural selection, and the plausibility of individual life forms creating other life forms. One could redact the particulars, the specific molecules or the dates, and this story would become an outline for how any given life form could create a revolutionary phenotype.

Toward the end of this book, I revisit this outline to present a thought experiment in which human DNA shares the fate of the bygone RNA queens—that is, a scenario in which human societies cause a phenotypic revolution which results in the complete destruction of DNA-based life on Earth.

Another reason why this story is much more plausible than any theory ever proposed concerning the origin of DNA is that every single step of it can occur over millions of years, without a single

THE REVOLUTIONARY PHENOTYPE

event requiring any other explanation than the basic theory of evolution. There is no point at which we have to invoke some sort of very-low-probability event, such as some lucky mix of atoms forming in some pond, or invoke some unlikely circumstance resulting from special meteorite collisions with the Earth. The story flows straight from the first principles of evolution—the idea that replicators tend to create machines that are increasingly servile to their reproduction. All of the steps above are an inevitable fatality if the revolutionary phenotype happens to be a better carrier of genes compared to the native replicator.

When looking at the entire story, it may seem like the principles of natural selection have been violated from the perspective of the RNA replicators because they created a machine that ended up working to their disadvantage—was there not a violation of the principle of phenotypic servitude when RNA's phenotypic machines ended up replacing RNA? No. It is only the story as a whole that violates the interests of the RNA replicators. Considering single steps one at a time leads to no violation of the first principles of evolution. It is simply the case that natural selection initially favored RNA replicators, eventually came to favor organisms that allowed the transfer of information between RNA and DNA, and finally ended up favoring organisms that used DNA replicators. In other words, while the principles of natural selection predict that phenotypic machines become more and more servile to their replicators over time, there are no rules that safeguard life forms from being outcompeted by the new replicator varieties they generate in the pursuit of their own interests.

Beyond the revolution

The RNA-world hypothesis implies that each modern DNA-based organism on Earth, including humans, can be considered as a special type of phenotypic machine, which I call an embedded phenotype:

Embedded phenotype

The phenotype of a replicator that has triumphed in at least one phenotypic revolution (that is, a replicator that used to be a phenotypic machine of another life form).

The fact that we are embedded phenotypes means we have ancestors in our lineage from which we are the revolutionary phenotypes rather than the genetic descendants. One of the shocking consequences of this realization is that we have to find a new way to draw genealogic trees, since there are now two ways we can be related to our ancestors:

1. by being their genetic descendant,
2. by being their revolutionary phenotype.

In Chapter 10, I propose a new way of drawing genealogic trees to account for this reality.

One of the details that deserve our attention in the storyline regarding phenotypic revolutions is that DNA was not satisfied with just discarding the RNA queens in order to build its new existence. Instead, DNA subjugated the RNA queens, exploiting them in nearly all physiological functions. The RNA life form was recycled *into* phenotypic machines that sustained and served DNA for over four billion years. Even in your own body and in the cells of all other DNA-based organisms on Earth, there is close to nothing that DNA does by itself. DNA almost always passes through RNA molecules to perform its work. One way to view these facts is to say that revolutionary phenotypes interact with the world by creating fake genes of the native replicator that has created them. In truth, revolutionary phenotypes always work this way, and the reason why will be explained in Part II.

If the extinct RNA queens could recount the insights gleaned from their demise during the phenotypic revolution of DNA, they might state the following:

We produced DNA, a machine that initially supported our replication;

The machine resembled us in many ways;

We started trusting this machine to intermittently carry

the genes of our offspring;

The machine seized our replication cycle;

Four billion years later, this darned machine is still using us to further its own replication. It has now neutralized most of our reproductive abilities.

From one perspective, RNA was master of its domain, kindly employing DNA in its toil until DNA became too unruly. From another viewpoint, DNA could not put its best foot forward under the boot of RNA, and, as a superior entity, it gradually outwitted RNA and escaped its shackles. In either formulation, when the phenotypic revolution was completed, DNA likely destroyed, crowded out, or otherwise suppressed the untamable RNA queens, those who still refused to stop replicating. This deathblow was quick and painless, coming after DNA had essentially wrested control of every biological function in the organism. DNA dispatched the RNA queens when their services were no longer required.

Today, RNA viruses may be the only remnants of the RNA queens of yore. I like to think of RNA viruses as the last scream of the RNA life form—a scream which has echoed for billions of years, yelling "We were here! We made you!"

If one were to depict the rise of DNA through a social media meme, it would be a picture of a smiling fellow with menacing eyes, saying to its RNA master:

I'm in your phenotype, eradicating your life form.

In the unlikely event that any life form on Earth ever starts truly caring about its own existence, it might pluck the following piece of wisdom from the ashes of the RNA world:

If you're one of the most successful replicators on Earth, do not invent a machine that resembles yourself.

Gaming the system

The combination of all humiliating defeats across computer gaming history could never compare to how mercilessly RNA was owned by DNA. No act of multicellular aggression, no siege, no conflict since has been as dramatic as the usurping actions of DNA: its total enslavement of another life form, and its recycling of it for its own interests for billions of years.

We know that evolution is rough. We know our ancestors, those who made our lives possible, were some of the most vicious beings who ever walked the earth. We know the reason we exist today is that our ancestors successfully raped, murdered, cannibalized and pillaged countless other human cultures. However, there is something even more deeply vicious in what DNA did to RNA.

First, DNA did not pillage RNA's house, village or food reserves; it pillaged RNA's very genetic material and reused it in its own creative ways. Second, the revolutionary phenotype, DNA, did not come from outside of the RNA organism. Rather, it started off as a piece of the organism it eventually hijacked. This would be like if a cancerous tumor, instead of just killing you, started building its own civilization within your body while keeping you alive just to buy some time until it could recycle the various pieces of you it needed to accomplish its greater purpose.

In the storyline of DNA's phenotypic revolution, I have left a major question unanswered: What function did DNA first serve for the RNA queens before it revolted? We can only make educated guesses at the moment. One recent hypothesis, proposed by Patrick Forterre[11], suggests that DNA was deployed as a virus to thwart the defenses of competing RNA-based organisms. While all the defenses of the organisms at that time would have been up and prepared against RNA invasions, some clever queen may have started to use DNA, a new molecule that could go unrecognized, as a Trojan horse to penetrate RNA-based organisms without triggering their defense systems.

Other hypotheses are just as valuable: DNA may have been a storage device that allowed the RNA queens to live longer without suffering the high rate of mutations that make RNA less chemically stable than DNA. Imagine being offered a technology that makes your genome and body survive up to a thousand years instead of the 75 years or so you can currently hope for. Perhaps that offer resembles what the RNA queens were confronted with when they developed DNA, a molecule that could remain stable for years. RNA paled in comparison, as it would mutate into uselessness after only a few weeks or months. Ultimately, whether DNA was travelling between organisms or strictly remained an internal device used within single organisms, the storyline of the phenotypic revolution described above remains the same.

Our understanding of the RNA world permits confidence in the above storyline as an outline for phenotypic revolutions, although some details remain unknown: Did the revolution occur within an hour or over a hundred million years? Once DNA had established a means of reliable self-replication, was it long before it deposed the outmoded RNA queens clinging to their RNA-copying ways? How many failed rebellions were there? Did DNA storm the RNA Bastille once or a billion times? Were there organisms with both RNA and DNA replicators, and did these organisms persist? If so, for how long? How much information was exchanged between RNA and DNA during the revolution? How many of the modern DNA genes do we owe to the RNA queens? Was the whole affair a skirmish between two replicators within a milliliter of ocean water, or was it a molecular world war?

Chapter 6

The Gene, The Meme and The Quene

How should we approach the idea of multiple replicators living within single organisms, as was the case during the phenotypic revolution of DNA? When two replicators live in the same organism, is it possible to identify one as being the phenotypic machine of the other? These questions are strikingly similar to some of the questions asked by thinkers in the field of memetics. Perhaps the most ambitious claims of memetics are those of Susan Blackmore. In *The Evolution of Meme Machines*[12], she writes:

> We should remember that this new kind of evolution proceeds not in the interest of the genes, nor in the interest of the individual who carries the memes, but in the interest of the memes themselves.

Similarly, in *The Meme Machine*[13], she writes:

> The theory starts only with one simple mechanism—the competition between memes to get into human brains and be passed on again. From this, it gives rise to explanations for such diverse phenomena as the evolution of the enormous human brain, the origins of language, our tendency to talk and think too much, human altruism, and the evolution of the Internet.

Should we, like Blackmore, credit the evolution of the human brain to culture itself rather than natural selection of the DNA genes that form it? Is this question semantic, or is it pointing to a valid, fundamental interrogation?

It is undeniable that culture modifies the environment in which genes evolve, and genes in turn modify human culture. At first glance, it may seem like we can arbitrarily view the issue from either perspective. This conclusion, however, would be premature until we can figure out answers to the following questions:

Is there any criterion by which we can conclude that, when two replicators reside within the same organism, one is evolving *for* the other, or are we bound to answer instead that "every element influences every element"? Is there a robust criterion by which we may say that, in the tango between genes and memes, one of the two is a replicator, while the other is its phenotypic machine? Similarly, when it comes to modern organisms, is there a criterion by which we may declare DNA a replicator and RNA its faithful phenotypic machine? More generally, when two replicators live together in the same organism, how do we determine which one is subject to natural selection?

Let us ponder once more the RNA life form and its demise wrought by the DNA machines it birthed by comparing the interaction to a tango. Asking whether natural selection applies more to memes or genes, or more to DNA or RNA, may be like judging which member of a tango duo is the superior dancer. There might be no right answer. The most stunning dancer may shine only through the incredible talent of their partner. Within the proper pair, even a klutz may glide effortlessly.

Evolution is not like dance, however, as human culture has evolved to serve genes. And it did so in ways that do not violate the first principles of evolution, from the perspective of DNA. As we will see in Part II, there is indeed an asymmetry between genes and memes, which leaves all of the evolutionary power in the hands of genes.

Liane Gabora points out in *The Origin and Evolution of Culture and Creativity*[14] that culture is grounded in biology just as biology is grounded in the physical properties of matter. Extending this reasoning, we can say that, just as the existence of DNA-based life on Earth violated no laws of physics, so too has human culture violated no laws of DNA-based genetic evolution.

Since we are exploring embedded, interacting life forms, a label for replicators within these systems is helpful. Susan Blackmore used the term "memes" to indicate cultural replicators and "tremes" to describe technological replicators (i.e., replicators that exist in the data structures of modern computers which make copies of themselves). In the long run, this is an untenable system as we would incrementally exhaust the alphabet with each hypothetical phenotypic revolution. As such, I use the word "quenes" (pronounced "queens"), as a general term to describe the minimum units of inheritance of any replicator. Memes, tremes, and genes, and whatever else you can imagine making a copy of itself are all types of quenes.

Quene

A quantum of inheritance. The discrete, irreducible unit of inheritance of a replicator capable of making copies of itself within at least a certain environment.

DNA is composed of strings of molecular letters, and at the extreme, we may consider each letter to be a quene. In line with memetics, we may also consider the possibility that human culture is composed of quenes—discrete, indivisible bits of cultural information, such as the letter 'e.'

This use of the word quene is novel, but the idea follows from Richard Dawkins' *The Selfish Gene*. In his book, Dawkins argues that, because segments of DNA can be broken, cut, and spliced with other segments of DNA, a whole DNA molecule does not always qualify as a replicator. In describing the fundamental unit of selection, he argues that, as the DNA segments become progressively smaller, they are proportionally less likely to be cut and reconnected via these processes. Thus, the smaller the segment, the more "selfish" it is over evolutionary time. The ultimate extension of this idea is that the true replicators are the quenes—the individual letters of DNA, which cannot be cut without losing their ability to replicate.

Another advantage of this term is its applicability to atypical life forms, even those with ill-defined or unknowable physical bases. The creatures described in this book may exist in molecular space, like us, or they may exist in manifold places: in virtual realms, on other planets, within diverse information structures of some alternate universe, or among combinations of all three. Like us, they may evolve in time and exist in space, or they may exist in time and evolve in space. Liberal definitions are the best way to state truths about things that one still ignores.

Oddly, the present use of "quene" is akin to the original meaning of "gene," in that it points to a fundamental unit of inheritance. The word "gene" was coined by Danish botanist Wilhelm Johannsen following his observations of the heritability of certain physical features in plants. Without knowing that DNA was the carrier of the genes, Wilhelm knew that some irreducible factor had to underlie the heritability of various phenotypic characteristics. Advances in molecular biology later enabled us to identify the individual phenotypic machines that extract information from DNA to produce RNA strands and proteins. In other words, we now understand the molecular mechanisms that explain why positions of seeds in the pods of a plant are inherited across generations, as Johannsen originally observed. The process that converts information from DNA to proteins is known as **gene expression**, and its discovery and investigation have led to a denaturation of the word "gene." Wikipedia[15], for instance, currently defines a gene as follows:

> In biology, a gene is a sequence of DNA or RNA that codes for a molecule that has a function.

There are two problems here:

First, the definition of genes was restricted to those pieces of DNA that code for molecules. In this book, though, we do not presume all parts of a replicator are leading to the production of functional molecules. A replicator is simply something that makes copies of itself, independently from whether or not it has a function or whether or not it leads to the production of a molecule.

Second, our term for the replicator must not be used for stretches of RNA, since they do not make copies of themselves. Thus, let us abandon the term "gene," as it has been soiled by the collective.

I have opted to employ the pithy "quene" in order to designate the elementary inheritance units of a replicator. Its consonance with the word "queen" is a random but fortunate occurrence, since it turns out that the queens of insect colonies are the organisms responsible for copying the genetic material to the next generation. In other words, the role of a queen in an insect colony is to carry and replicate its DNA quenes, while the sterile workers of the colony are the phenotypic machines of the queen (and, also, of the quenes).

PART II

The Answers

Chapter 7

The Forgetful Qreamplex

In Part I, we explored the mechanism by which phenotypes, across generations, become more and more servile to their replicators. Simply put, the replicators able to create good phenotypic machines replace other replicators by outnumbering them. If you extrapolate this principle over millions or billions of generations, you can understand why a bacteria could progressively evolve into a bird. All innovations that help survival and reproduction are kept and even amplified by natural selection, while those that impede reproduction are progressively abandoned. But how can we distinguish phenotypes from replicators in the first place?

There are three main characteristics that differentiate phenotypes from replicators. Identifying these characteristics leads to an understanding of what a phenotypic machine must overcome before succeeding in a phenotypic revolution. In this chapter, we will identify the first of those characteristics—one that affects only some phenotypic machines. I call it forgetfulness.

Let's explore the problem of forgetfulness by considering hemoglobin, the protein that helps blood efficiently distribute oxygen throughout our bodies. The key functional component of hemoglobin is called the heme, which is a phenotypic machine that carries in its center a critical ion of iron. The individual atom of iron carried by each heme has a profound and felicitous affinity for oxygen, which in turn is vital for energy production in all animals. Thanks to the heme, hemoglobin is capable of delivering oxygen from your lungs, where it grabs the atoms of oxygen, to all other parts of your body, where it drops those atoms. And because hemoglobin packs more oxygen into a smaller amount of liquid, this oxygen-delivery system is far more efficient than one that passively diffuses oxygen without hemoglobin.

Hemoglobin is just one of countless phenotypic machines produced by our DNA. If we were to trace the lineage of a single hemoglobin gene in your genome, we could trace it back through your parents, their parents, your ancestors, all the way back to some ambitious, prehistoric fish. Your ancestors had sequences of DNA that eventually led to the production of hemoglobin, and you inherited those DNA quenes. Those who inherited dysfunctional or different versions of those DNA sequences have probably died without reaching the age of reproduction.

At a specific point during your embryonic development, some cells transformed into factories that would produce red blood cells which are in turn responsible for carrying hemoglobin across your body. The information necessary to produce hemoglobin was read directly from your DNA by proteins specialized in that task. These transcription proteins approached your DNA, honing in on special sequences of DNA quenes that were interpreted as "go" signals, which told them where to begin reading the information. They then began transcribing RNA molecules based on the blueprint they were given by the DNA molecule. These RNA molecules were then translated into a massive number of hemoglobin proteins. Once fully assembled, the completed hemoglobin proteins were ready to function as an efficient, oxygen-carrying machine in your red blood cells.

There are two faces to hemoglobin. On the one hand, hemoglobin proteins are phenotypic machines. On the other hand, the sequences needed to produce these proteins are located in your DNA. These DNA segments are the replicators—the entities that are transmitted from one generation to the next. We can tell the difference between a replicator and a phenotypic machine using the following thought experiment:

Suppose I take all the hemoglobin out of your body and make a small revision to the amino acid sequences that constitute those proteins. Suppose the change I make has a minute impact on the function of your hemoglobin proteins, and I very quickly return those edited proteins to your blood (so quickly that you feel nothing and keep on living your life as usual). After I make this change, if you reproduce, your kids will not inherit the modified version of the hemoglobin. Why? Because hemoglobin proteins do not make copies of themselves, and a modification to the protein is not perpetuated across generations. Therefore, your genome will not reflect my changes. That's why the hemoglobin protein isn't a replicator—it is forgetful.

Now let's tweak this example slightly. Suppose that, instead of altering all of the hemoglobin proteins in your body, I simply modify all segments of DNA that encode hemoglobin with a similar, inconsequential change. In this case, the letters of DNA I altered would not only ensure that every future hemoglobin protein produced in your body would bear my revisions, it would also ensure that a share of your future offspring and descendants would produce modified hemoglobin proteins. (As a side note, some of your offspring would not produce the modified version because they would inherit the unmodified gene from your sexual partner rather than from you.) That's why the hemoglobin gene, the sequence of DNA quenes that encodes the hemoglobin protein, is a replicator—it is not forgetful, and it leaves a copy of itself in subsequent generations.

The hemoglobin protein is not the only piece of our phenotype that cannot copy itself from one generation to the next. Most of our phenotype can't! Throughout our lifetimes, our bodies are battered by the environment: burned by hot surfaces, bitten by wild animals, and pummeled by a terrifying array of pathogens. While these events may scar or even destroy parts of our bodies, they typically do not change our DNA. Hence, our injuries do not show up in our descendants. If you have a scratch on your arm, your kid will not inherit that scratch. He may only inherit your tendency to get scratches, leading him to acquire analogous scratches of his own.

This is forgetfulness in action. The title of replicator is not appropriate for the scratches on our body precisely because they do not replicate, nor do Mars-rover prototypes or the one-of-a-kind handwritten birthday cards you made for your parents when you were eight years old. Unlike DNA, these things do not benefit from what we will later define as a "printer." In other words, these things lack a reliable read/write mechanism as part of a reproductive cycle.

I refer to such elements as qream:

Qream

The ensemble of all things in the universe that do not make copies of themselves. This includes every single entity that is forgetful, i.e., any entity that will never be directly read and copied by anything, or that finds itself in an environment where such a copying process is impossible.

Perhaps the consonance with "cream" will be worrying to some readers, and critics may claim that my reductionist approach fails to recognize the inherent complexity of all the non-reproductive matter in the universe by reducing it to "qream." Hence, let's call all the things that do not make copies of themselves the **qreamplex**, in recognition of their potential complexity. The qreamplex includes the detailed, grain-by-grain arrangement of the asphalt covering your local highway, the scratches on your body that your kid will not and cannot inherit, and the stifled ideas in your brain that never get communicated, among so much else.

As we have indicated above, the qreamplex *does* include things that resemble each other, which may sometimes fool someone into believe that a copying process has been in operation. But do not mistake these for replicators. Multiple copies of a thing may be present in the qreamplex, but some have not been produced through a process of replication. The hemoglobin proteins in your blood do resemble the hemoglobin in your child's blood, and it is easy to imagine that this is due to some copying mechanism. However, what allowed the transmission of the hemoglobin protein is not a process of replication of the protein itself; it is the replication of DNA. In contrast to phenotypic machines, replicators do not rely on other replicators to make copies of themselves; they are the entities that cause their own copying. While phenotypic machines do show up in multiple copies, there is nothing that reads and makes a copy of them.

In order to be considered a quene, the multiple copies must have been produced through printing. When a phenotypic machine shows up in multiple copies without having been printed, we will say that its copies are "rooted" in a replicator, using the following definition:

Root

The set of quenes in a replicator that are responsible for causing the production of multiple copies of a given phenotypic machine within or across generations. *Ex: the hemoglobin proteins are rooted in the sequences of DNA quenes that we call the hemoglobin genes.*

In short, when multiple copies of an entity keep showing up across generations within living organisms, there are two possible explanations:

1. The copy of the object is a direct copy of the other, in which case it is a replicator—a set of quenes;

2. The two objects were caused by a similar chain of physical phenomena emerging from another entity which was itself a replicator, in which case they are phenotypic machines rooted in a set of quenes.

Lessons from the phenotypic revolution of DNA

As mentioned earlier, to instigate a phenotypic revolution, phenotypic machines must overcome three problems, the first of which is the problem of forgetfulness:

The problem of forgetfulness (#1)

Changes that occur to phenotypic machines are not preserved in the next generation. Only changes that occur to their replicator will be transmitted to offspring. Therefore, the principles of natural selection do not apply to phenotypic machines in the same way they apply to replicators. Phenotypic machines are unable to improve across generations in a way that would favor their own replication, other than by serving the replication of the replicators in which they are rooted.

While forgetfulness is a big problem for the phenotype, it doesn't seem to have plagued DNA. How exactly did DNA solve the problem of forgetfulness? The answer lies in the details of the phenotypic revolution of DNA presented in Chapter 5.

DNA was not like all other phenotypic machines; it was not merely released into the world in the hope that it would contribute to the replication of the RNA queens, although that might have been its initial purpose. The RNA queens eventually used DNA as temporary, external storage media for their genes. As a result, the mutations that occurred within DNA were transmitted to the subsequent generation as the DNA molecule was the blueprint used in the production of the RNA queens' offspring. This illustrates a solution to the problem of forgetfulness, which can be expressed as follows:

Solution to Problem #1

Revolutionary phenotypes solve the problem of forgetfulness by encoding their own states into the next generation of offspring by using the system of genetic inheritance that already existed in the organisms of the native replicator.

The moment one RNA organism found a way to store its genes into DNA was also the moment the DNA machines found a way to store their mutations into the RNA quenes. I previously referred to this interaction between RNA and DNA as a tango. Now may be the right time to formalize this concept:

Replicator tango

An arrangement in which the genes of a native life form are stored in an external device, which is then trusted to determine the genetic makeup of the next generation of offspring of the native life form. Replicator tangos occur when organisms store their genes in two separate physical media rather than only one.

The concept of a replicator tango solves the problem of forgetfulness as it involves a phenotypic machine using the genetic transmission system of its native replicator to store its own information. Once a replicator tango is established, the distinction between phenotypic machine and replicator completely dissolves as far as the two molecules are concerned. In fact, both molecules in the tango can be equally seen as replicators as they both use each other to send information to the next generation. In other words, they are rooted in one another.

Is there any solid evidence that a replicator tango existed between DNA and RNA at some point on Earth? Absolutely. In one particularly relevant study, Leipe and colleagues[16] surveyed the three major branches of life on Earth: bacteria, eukaryotes (which includes human beings), and archea. They reached the fascinating conclusion that the machinery underlying DNA replication in these modern forms of life differs vastly between bacteria and the two other groups, eukaryotes and archaea.

The machinery for DNA replication in these organisms is dissimilar enough to suggest that DNA replication may have two independent origins within the tree of life. On the one hand, all three of these groups use DNA, as well as a shared assortment of cellular machinery, so they must share a common ancestor. On the other hand, the DNA replication mechanisms are so divergent that it is very likely they blinked into existence separately.

A parsimonious solution to this enigma is to propose that the common ancestor we share with bacteria was not a DNA-replicating organism at all. Rather, it may have been an organism governed by a genetic code embroiled in a replicator tango between RNA and DNA. The completion of the phenotypic revolution—the moment at which DNA first self-replicated and hence disengaged from the replicator tango—may have occurred twice, independently, spawning bacteria on one branch and archaea and eukaryotes on the other. If that is the case, we would share common ancestry with bacteria but not a common revolution, and the yoke of RNA's oppression would have been thrown off not once, but twice!

Further evidence of ancient replicator tangos can be found in the mechanisms used by viruses to replicate. Some viruses are DNA-replicators, other viruses are RNA-replicators, but some other viruses might be called "tango replicators," because they pass through stages in which they alternately encode their genomes in DNA or RNA. Accordingly, to anyone who doubts the existence of replicator tangos between DNA and RNA, we might say, "Well, actually, they still exist and might be occurring in your body at this very moment."

In conclusion, we have found that replicator tangos are a method that revolutionary phenotypes can use to pass on genetic information to the next generation, thereby solving the problem of forgetfulness. They do so by essentially hijacking the genetic transmission system of the life form they are usurping, using that very system to transmit their own genetic information.

We now move on to the second problem faced by any phenotypic machine attempting to start its own life form: How can a phenotypic machine survive without the organisms produced by its creators?

Chapter 8

The Naked Warrior

Imagine a warrior on a battlefield, equipped with an armor, a sword, and a shield and surrounded by his fellow soldiers. Using the training he has learned from his masters and all of his resources, he effectively slashes through his opponents, cutting them to ribbons and easily gaining ground. Now imagine that we suddenly remove all of the warrior's equipment, all of his clothes and armaments, and even the army that accompanies him. In other words, imagine a solitary, naked warrior standing against an army of thousands, performing the same actions he was trained to do, just as if he was equipped for battle. Suddenly, the movements that made sense when the warrior was well-equipped do not make sense anymore. His thrusts and parries become completely ineffective because he no longer wields a sword. Raising an elbow no longer protects him from attacks because he has no shield. Summoning help from his fellow soldiers becomes ineffective because no one is there to listen.

The naked warrior's problem is quite similar to the problem that would be faced by a phenotypic machine that attempted to live and replicate on its own. Even if a phenotypic machine succeeded at solving the problem of forgetfulness, it would still face great difficulty from being isolated from the organism in which it started to replicate. Imagine, for instance, a brain that would start to make copies of itself. The copies of the brain would not survive without blood vessels, a heart, lungs, and a cranium. The problem is that the key to constructing all of these crucial organs is found exclusively within DNA.

The naked warrior problem (#2)

Phenotypic machines typically have functions only in the context of the organisms they find themselves in. These organisms typically are constructed by the replicator. In the absence of the environmental context created by their replicator and by the other organs with which they have co-evolved, most phenotypic machines would instantaneously die, or at least lose some or all of their functions. This makes it unlikely that even those phenotypic machines that do make copies of themselves would succeed at surviving on their own.

Phenotypic machines face the problem of the naked warrior every day. In your skin, for instance, there are cells that are able to make copies of themselves. When your skin is wounded, the multiplication of these cells is an important part of the process that allows your skin to heal. However, if you were to cut off a patch of your skin and leave it on a counter, the cells that normally work to repair injuries in your body would not be able to reconstruct a healthy patch of skin, much less a full body to live in. That's because the patch of skin left on its own has not evolved the series of tricks that allow for its proper reconstruction. It needs the blood vessels that transport oxygen and nutrients. It needs the neuronal wires that carry sensory signals back to the brain. In the absence of its normal environmental context, the patch of skin is very unlikely to survive, and in fact, it will have difficulty accomplishing any of the functions it would normally execute flawlessly in a proper body.

Sterile worker ants face a similar situation in ant colonies when their queen dies. Suddenly without a colony to report to, the solitary, sterile worker will typically continue performing some of its tasks. It might still carry some pieces of food, or it might participate in the digging of tunnels. But this lost, sterile worker is a naked warrior. In the absence of the rest of the colony, these activities are completely meaningless as they do not serve the reproductive success of the queen. It is only a matter of time before the sterile worker dies without leaving any descendant (other than those that had already been produced by the queen).

Let us go back to the heme, the piece of hemoglobin that carries oxygen. Imagine for the sake of argument that a heme has solved the problem of forgetfulness by finding some way to make copies of itself. What would happen if such a self-replicating heme was to become independent from DNA? Suppose the organism carrying this heme died. The heme could keep replicating from inside the dead body of its host, but for how long? Soon the sugar necessary to generate energy would run out. The heart would stop beating, making it more difficult for the self-replicating heme to survive and travel through the body. Eventually, the dead body would succumb to various flesh-eating bacteria, and the heme would be confronted by a much more hostile environment than anything it ever had to face when it was peacefully residing within the boundary of a well-protected, living human body. The few generations that the self-replicating heme would be able to generate in these conditions would be insufficient to allow it to evolve solutions to these numerous problems. The human body had four billion years to evolve such solutions. In contrast, our hypothetical self-replicating heme would only have a few hours after the death of its host to re-invent everything.

The naked warrior problem makes it even less likely that a phenotypic revolution would be successful because, even when a phenotypic machine acquires self-replication, it is unlikely that this phenotypic machine will be able to survive on its own without the rest of the phenotype. This is why cancer cells, while they are self-replicating and have devastating consequences on single lives, are not to be worried about in terms of becoming their own life form—they die with the body they kill.

We now have to repeat our thought experiment of Chapter 7 and ask how DNA solved the problem of the naked warrior during its own phenotypic revolution.

Interestingly, the idea of replicator tangos solves the naked warrior problem, just like it solved the problem of forgetfulness. By engaging in a replicator tango, not only does a revolutionary phenotype evolve a way to create true copies of itself in the next generation, it also adapts a method to construct novel organisms in which it can live. Replicator tangos do not compromise the integrity of the original quenes of the native life form, since they reproduce these quenes as well.

The solution DNA found to the naked warrior problem can thus be summarized as follows:

Solution to Problem #2

Revolutionary phenotypes solve the naked warrior problem by forming a replicator tango with the native life form. They then use their capacity to encode the native life form's quenes to produce organisms in which they can live and continue to evolve.

At this point, we have identified two problems that were solved by DNA during its phenotypic revolution—problems that have kept most other phenotypic machines from becoming life forms of their own:

1. A phenotypic machine cannot be subjected to natural selection as long as it is forgetful (Problem #1);

2. A phenotypic machine is unlikely to survive outside of the organism that has produced it (Problem #2).

These first two problems explain why, despite the fact that your body contains more than 30 trillion cells, not a single one of them has ever started a life form of its own. In fact, none of the 30 trillion cells carried by each of your millions of human ancestors have ever found a way to survive on their own. If one had, we would find single-cell organisms that happen to carry the human DNA genome. The fact that not a single one of these cells happened to have some accident that would allow it to survive is a demonstration of the strength of the first two problems.

However, Richard Dawkins himself could see that these two problems led to an impasse: either the list of problems is incomplete, and there is some third, undiscovered problem for the phenotype, or there are some phenotypic machines that are indeed subject to natural selection, like replicators. The prime example given by Richard Dawkins in support of the latter conclusion is human culture—as we have pointed out before, memes do make copies of themselves. They are not part of the qreamplex.

Were it just for these two problems, it would seem that human culture, or memes, can indeed be characterized as an independent replicator, just as Richard Dawkins intimated. Memes have solved the problem of forgetfulness in that the useful information they carry gets preserved as they copy themselves from brain to brain. Similarly, memes have solved the problem of the naked warrior because they jump from one organism to the other. As long as there are new brains to invade, they will not face the problem of the naked warrior. And up to now, we have not identified a criterion that would keep memes from being treated as a replicator subject to natural selection, leaving us stuck at the same point as every thinker in memetics over the last few decades.

The next chapter identifies that missing criterion, which we will call the third problem of phenotypic machines.

Chapter 9

Trickster Printers

Our first principles of evolution have already left space open for the idea of mutations. Mutations are imperfections in the copies that a replicator makes of itself. The theory of natural selection relies on these mutations. If the copies of a replicator were always perfect, the replicators would not evolve because there would be no difference in phenotypes across the population and thus no variation that would allow natural selection to determine the success or failure of a particular phenotype. Everyone would be equally successful (or unsuccessful).

Up to now, we have assumed that replicators were located in an environment where some sort of high fidelity printer could simply scan the replicators and produce their (almost-perfect) copies. In this chapter, though, we will consider situations where the printer is a more complex entity, capable of cheating during that copying process.

First, let us clarify exactly what we mean by printer by using the following definition:

> **Printer:** The set of physical entities that carry out the copying of the replicators' information in the process of creating a new generation of replicators. This includes any entity that causes mutations to appear or not to appear during the copying process, thus making the new replicator less or more dissimilar to the original version. Specifically, the term printer includes any entity that controls the content of the replicator, but not those that merely select replicators. For instance, if you have a strand of DNA in a mouse, eagles do not qualify as

printers because all they can do is eat the DNA strand or let it live. This is selection. The term printer is limited to those entities that have access to **determining the content** of the replication through mutations. By this definition, an office printer qualifies as a printer because it can in principle determine the content of the copy and alter that content. However, a fire affecting a part of a library, reducing some books to ash while leaving others intact, is an event of selection, not printing, because the fire has no way to change the content inside the books.

In the case of DNA-based organisms, the printer is a molecule called DNA replicase, along with a set of other molecules which participate in various operations related to reproduction, such as those molecules that cut and re-splice segments of DNA. It also includes error-checking molecules which activate after the copy has been produced and which may alter an erroneous quene back to its original state. In fact, the DNA printer includes the molecule of DNA being copied itself, since it is necessary that this molecule be present for the new copy to be generated. Together, these molecules and the laws of physics that apply to them determine whether or not a given quene of DNA will be correctly copied to the next generation. The DNA printer operates at very high fidelity. It makes about one error for every 10,000,000,000 letters of DNA it copies correctly (when taking into account repair mechanisms).

One thing must be pointed out concerning the operation of the DNA printer: the copying activities of these molecules are not controlled by any other form of complex, intelligent, or evolving entity—nothing other than the very DNA genes that participate in the production of the printer. These molecules do not seem to serve any other function than making the best copy of DNA they can. This is different from, for instance, a newspaper editor, who while reviewing an article, will change the information according to certain content preferences. Perhaps some sentences will be removed to comply with commitments to advertisers. Perhaps some phrases will be changed according to political preferences. In contrast, the DNA printer seems to care only about making a copy that's as good as it can be, given its physical constraints. The few copying errors that occur during this process are not initiated with any goal in mind, nor are they linked to any other evolutionary interest than the survival and reproduction of the DNA strand being copied. DNA mutations seem to appear at random.

But random how? To a certain extent, true randomness does not exist in our physical universe, at least as far as classical mechanics go. We live in a deterministic universe in which the bumping of atoms into one another follows certain laws and respects the principles of causality. Therefore, nothing is truly random. There is always a physical explanation for why something happens, at least at the macroscopic level, which is what we are concerned with.

Imagine we are sitting at a casino poker table, watching a dealer deal out cards. We assume the cards, if they have been properly shuffled, are coming in a random order. However, we also recognize that, given an appropriate computer-generated model that includes the composition of the deck of cards, the hands of the dealer, and the physical collisions of the cards during shuffling, it could, in principle, be possible to predict exactly which cards are coming next. In that sense, randomness is always relative to the observer and to their ability to perceive the physical mechanisms at play in the generation of a particular outcome.

When I say that DNA mutations occur on a random basis, I mean they are caused by physical processes that are not goal-oriented, intelligent, or evolutionary, and that are more or less unpredictable, at least within the scope of biology.

Even so, it is easy to conceive of printers that would vastly differ from the DNA printer. Imagine an office printer for instance, which instead of faithfully copying your documents, would have preferences about what those copies contain. Every time the printer saw the letter "w," it changed it to "n." Or every time it saw the word "socialist," it changed it to "capitalist." Because we are interested in outlining the evolutionary interactions in such systems, we will concern ourselves only with the ultimate causes of such defective printers when they are rooted in a replicator. What kind of trickster would program such a printer?

The trickster who made the printer change the word "socialist" to "capitalist" might have been motivated by political interests—interests that likely had an emotional basis. This emotional basis was most likely encoded in the trickster's brain. But why were such emotions motivating the trickster in the first place? The answer is that its brain was a phenotypic machine of DNA. Ultimately, the cause of the switch from "socialist" to "capitalist" must be, to some extent, due to the DNA of the trickster, because DNA is part of the physical chain of causality that leads to all human behavior. Thus we will define such manipulated printers as follows:

Trickster printers

Printers that introduce mutations into the replicator they copy, and where the mutations are rooted in a separate replicator (i.e. the mutations are the phenotype of another replicator). We call the separate replicator at the source of the mutations the **trickster** and the replicator being copied and/or modified the **fool replicator**.

In the face of trickster printers, are fool replicators still able to evolve according to natural selection? Can a life form improve itself when another life form is in charge of ensuring its faithful copying?

Let us first consider the simplest case, that is, a set of fool replicators that are mutating and whose mutations are not affecting the survivability of the trickster replicator. In this instance, it appears that the fool replicators are indeed capable of evolving according to a limited form of natural selection as far as their comparative advantages go.

In order to illustrate this possibility, consider the following hypothetical scenario. Imagine a social media meme that has no consequence whatsoever on the human ability to survive and produce offspring. The social media meme spreads across the population and reaches some degree of popularity. Then, an unexpected error occurs in the process of transmission that changes the meme in a way that makes it even funnier and appealing to a larger segment of the population, again without affecting their ability to produce offspring.

The modified meme ends up spreading even more because it is funnier and more interesting than its ancestor. Here, we have a case where memes are competing with one another and replicating according to their own relative advantage, though not interfering with the audience's ability to produce offspring.

The above paragraph is a theoretical example that assumes these evolving memes would not influence the population's fitness. But one is left to wonder how much of human culture could be explained by the mechanisms presented in this artificial scenario. Most of what humans talk about on a daily basis has the potential to affect their reproductive ability: the kind words we tell our sexual partners, the opinion piece we publish in newspapers with the aim of affecting the political order of our tribe, the teaching of a proper alphabet to our children so that they are able to interact with people in their surroundings and eventually get a job, find a mate, etc. Even the information we share just for the laughs may have severe consequences on whether or not potential sexual partners will be attracted to us. A clever joke can get you laid, while a misplaced one can result in a slap to the face. One may wonder if a single instance of human communication has ever qualified as completely impactless on human reproduction in the history of our life form.

A more realistic scenario is to consider what happens when the fool replicator's actions do have the potential to affect the evolutionary fitness of the trickster. Thus let us modify our previous social media meme example.

Suppose we have a series of memes competing with each other on social media. Each meme is being copied by a human in the form of retweets and shares on Facebook and Twitter, and the websites allow humans to change the content of the sentence being shared. In this scenario, human DNA is the trickster. The human brain involved in copying, changing and sharing the memes is the trickster printer, and the social media memes are the fool replicators. One of the social media memes is a sentence that reads as follows:

Make moral babies.

For simplicity, let us suppose that any human who shares this sentence is somehow more likely to invest a considerable amount of time educating their children in a solid moral framework so they become decent people. Let us also assume that, as a consequence of this time investment, these people, overall, produce less offspring and therefore reduce their evolutionary fitness.

Nevertheless, the moral babies meme spreads with some degree of success, and in fact is passed on for thousands of years, thus lowering the reproductive success of everyone who shares it until...

A mutation occurs in one human, making him incapable of passing on that meme. This particular human being is the trickster. When passing on any meme, the trickster first parses the meme and replaces all instances of the letters "al" with the letter "e."

The trickster being exposed to an incoming flow of moral baby memes therefore systematically changes the meme before sharing it, spreading the following sentence across his social media:

Make more babies.

For simplicity again, let us suppose that any human who shares this new sentence is somehow made more likely to produce offspring, and therefore increase their evolutionary fitness.

According to simple math, in this system, as long as there are people sharing the moral baby memes, there will be a counter-evolutionary force increasing the evolutionary success of tricksters. And in this fictional environment, there is one certainty: the tricksters are increasing their representation in the population, while the other people are headed toward extinction. Sooner or later, no one will even be capable of passing on any meme with the letters "al."

The fictional scenario of the moral baby meme illustrates what happens when a life form (in this case, a social media meme) relies on a trickster printer: its evolution is bound to occur only as long as it does not violate the evolutionary interests of the trickster. As soon as the fool replicator starts impeding on the evolutionary interests of the trickster, the trickster can counter-evolve, modifying its own "tricks," that is, inserting genetic mutations into the fool replicator so that it stops committing such violations.

Our theoretical example is a simple illustration of what must have gone on for hundreds of thousands of years of human evolution. Genes evolved emotional- and language-processing machines that, instead of strictly copying memes, would modify those memes and their impact on human behavior in order to distill the most useful information while discarding the rest.

We have thus obtained the following rule, which we can add to our theory of natural selection:

Addendum to the theory of natural selection

When a fool replicator is being copied by a trickster printer, the theory of natural selection only applies to the extent that the mutations of the fool replicator have no effect on the trickster replicator. For all other cases, the trickster replicator will win any evolutionary fight as it has the ability to modify the content of the fool replicator. Thus, any modifications of the fool replicator either favors the reproduction of the trickster to begin with, or if it doesn't, it imposes a counter-evolutionary force that modifies the trickster until they become apt at printing the fool replicator in a way that does not violate the trickster's interests. When its evolutionary fitness is challenged, the trickster can simply add mutations into the fool replicator until it becomes favorable again to the reproduction of the trickster.

Our addendum to the theory of natural selection describes the evolutionary relationship that exists within quenes of embedded life forms. In our original formulation of the theory of natural selection, we failed to mention that multiple sources could generate mutations. We now acknowledge that there can be two sources for mutations: they can either be random (i.e. intractable), or they can find their origin in a trickster printer, that is, an entity that is rooted in the evolutionary interests of another replicator. When this second situation occurs, the mutations affecting the fool replicator can be viewed as biased in that they favor the replication of the trickster. In such a scenario, the fool replicators are not allowed to explore mutations that would otherwise favor the fool replicators over the trickster. If they ever dared challenge the evolutionary fitness of the trickster, they would either fail because the trickster would counter-evolve a defense mechanism, or they would succeed and kill the trickster while dying with it. This is not the first time we have encountered this principle—as we mentioned earlier, cancerous tumors die with the body they kill. As it turns out, fool replicators also die with their trickster printer when they attempt to interfere with their reproduction.

We can summarize the addendum to our theory of natural selection with the following principle:

The principle of mutational servitude

When a trickster printer prints a fool replicator, the mutations occurring to the fool replicator are phenotypic machines of the trickster, and therefore the fool replicator is part of the phenotype of the trickster, and it evolves to serve the replication of the trickster.

Thus we can see that a phenotypic machine that would happen to be a fool replicator would be unable to evolve on its own:

The trickster-printer problem (#3)

Any phenotypic machine that solves the first two problems will likely still live within the vicinity of its native replicator. As such, it will likely require some machines of the native life form in order to perform its replication and mutation process. Thus phenotypic machines that solve the first and second problems are still likely to end up as fool replicators evolving under the evolutionary interests of a trickster printer. Because the trickster replicator has full access to rewriting the fool replicators, any evolutionary path that leads the fool replicator to impede on the fitness of the trickster can be met by a counter-evolutionary response, where the trickster simply modifies the way it copies the fool replicator until the trickster's reproductive interests are satisfied. This evolution will be determined by the trickster's interests rather than the fool replicator's interests.

We now have a better criterion to evaluate the relationship between various quenes that live as phenotypic machines within the phenotype of other quenes. We can use this new criterion to address the problem of memetics by simply asking where the trickster printer is in the relationship between human genes and memes. In order to determine which entity, between memes and genes, is the true victim of a trickster printer, one simply has to ask these two questions:

1. Do DNA quenes have access to modifying the quenes of human culture? Answer: Yes. DNA can produce different brains that will modify the culture they pass on in various ways. Everything in human culture, from single letters to syllable to the entire meaning of a sentence, and even the emotional meaning of an entire story, can be modified by a human brain.

2. Do the quenes of human culture have access to modifying DNA quenes? Answer: No. At the moment of writing this book, no human culture on Earth has reached a point where it determines which letters of DNA are to be modified in a human baby.

Thus, in the tango between cultural memes and DNA genes, the only entities that are the victims of a trickster printer are the cultural memes. There is a panoply of examples of trickster printing of human culture. The fact that we are able to discard certain details of a story, improve an existing theory, deceive others, or have social taboos that keep us from sharing certain sensitive ideas—all of these are signs that human DNA has been evolving as a trickster printer of human culture.

On the other hand, there is simply no cultural process that has the ability to mutate DNA in a way that would allow us to state that human culture is a trickster printer of genes. There are no social clubs that select the most desirable DNA gene mutations for the future generation of human babies. There are no countries that have outlawed specific naturally occurring DNA mutations. There is no à-la-carte service allowing parents-to-be to print the genome of their babies. As we will see in Chapter 12, we may be very close to letting human culture play with human genetics in this way, but we are simply not there yet, and until we are, we must conclude that memes are the fool replicators of a trickster printer—the human brain, produced by human DNA—and thus memes are not subject to a complete form of the theory of natural selection.

One point of contention may be raised concerning the distinction between printing and selection, a concept upon which our definitions rely. One anonymous member of my YouTube audience who goes by the pseudonym Fath'am has expressed this point of contention as follows:

> The problem with the idea of trickster printers is that the distinction between the act of printing genes and the act of selection is artificial, as both methods can lead to the same result. For instance, I could generate billions of bacteria, each having random alterations of some quenes of DNA, until I obtain the set of quenes that I'm interested in. In so doing, I would have essentially printed my desired genome without performing an act of printing, merely through selection.

This is an important critique which lays out that, in principle, any sequence of DNA obtained by printing an arbitrary sequence of quenes could also be obtained by the properly organized sequence of selection events.

In most cases in nature, selection events are nothing like printing, because a printer, as we defined it, has direct access to causing mutations in a replicator on a quene-by-quene basis, whereas selection events generally only have the ability to preferentially let certain replicators live and annihilate others.

Thus the distinction between selection and printing remains for most natural events. Indeed, outside of a human laboratory, we have never observed a life form that would be able to sequence the genome of other creatures and apply the kind of selection pressure strategy mentioned in the contention. However, we still must address the artificial case of the contention.

It turns out that the human who would engage in generating billions of bacteria and who would be selecting the bacteria in order to indirectly print a desired genome, would himself qualify as a printer according to the definition we have already provided. In our definition of printers, we have not excluded the idea that a process of selection could or could not be used as part as the mechanism for printing. The definition of a printer is based around the ability to cause mutations in the sequence of quenes in a replicator. How the printer does it is irrelevant to the definition. Therefore, we addressed the contention by highlighting that selection, although it is not necessary, can be one of the strategies used in order to perform printing.

Another contention comes from reasoning previously expressed by Liane Gabora[13], although these considerations predate the argument we are making here. Paraphrasing the arguments of Liane Gabora to adjust them to the current context, the contention would be that there is a continuum of possible effects that fool replicators could have on trickster replicators and that the sum of these effects may blur the line between selection and printing. We have previously indicated that, for instance, an office printer qualified as a printer, but a fire occurring in a library is a selection event. One could easily imagine a progressively more specific case of fire. Someone could align thousands of electronically controlled candles that would burn certain letters in certain books. At what point would the fire switch from being considered a selection event to being a printing event? First, let us define exactly what we mean when differentiating between events of selection and printing:

> The difference between an act of selection and an act of printing is ultimately one of precision, i.e., the degree of

precision of possible actions over a replicator. The first dimension of the impact that an event can have on a replicator is what we will refer to as **extent-of-access**. This means which region of a chain of replicators can be targeted by the event. The second dimension of the impact that an event can have on a replicator is **resolution-of-access**. This refers to the precision of the modifications that can be introduced to a replicator. Resolution-of-access can vary between entities that only have the possibility to alter large chunks of replicators (low resolution), down to entities that have the ability to change quenes on a 1-by-1 basis (high resolution). These two continuum can be thought of as a two-dimensional space. By printers, we mean entities that are very high on both the extent-of-access and resolution-of-access axes. In other words, perfect printers are entities that are capable of changing any quene of a given replicator, wherever they are in the genome.

It is beyond the scope of this book to try to pinpoint what happens to embedded life forms that would find themselves as intermediary points along the continuum between selection and printing. The reason we can avoid discussing these details is that, in all cases we observe in nature, entities find themselves clearly at either ends of the two-axes continuum. Consider the following examples:

1. Entities that are **clearly selective** in the sense that they have limited extent-of-access to the replicators and limited resolution on the replicators:

   A. A law in a country that disadvantages some people based on some physical characteristic. Such a law could only reduce the frequency of a specific set of genes in the population.

   B. A bird that is more likely to eat prey of a certain color. Such a bird would only reduce the frequency of a specific set of genes in the prey population.

C. A virus that kills organisms that do not have the proper immune system defenses. Such a virus would only destroy organisms that do not have a specific set of genes for immunity to this virus.

2. Entities that are **clearly printers** in the sense that they have full extent-of-access to the genome and full resolution as to their effect on the genome:

A. A replicator tango between DNA and RNA, whereby each quene of DNA is directly copied to a corresponding quene of RNA. Such a configuration allows the DNA strand to determine the entire content of the RNA strand.

B. A human brain transmitting a sentence of human speech. Such a human brain has the ability to alter the sentence—to completely change its meaning, syllable-by-syllable, or even letter-by-letter.

Thus, all of the cases that are relevant for this book are clearly at one of the extremities of the continuum between selection and printing. Ultimately those entities that we call trickster printers are entities that have high resolution and extensive access to the fool replicator.

For those who wonder about the precise mathematics describing when, exactly, a trickster can be said to have gained control over a fool replicator, I would point to the fact that the answer will depend on the frequency of random mutations of the fool replicator, the number of generations needed for selection to occur on those mutations, and the amount of mutations that can be attributed to changes by the trickster. Ultimately, as the trickster gains increasing mutational effects compared to the combined effect of random mutations and their selection within the fool replicator, it becomes progressively more unlikely for the fool replicator to take an evolutionary direction of its own. As indicated before, these considerations are irrelevant to cases occurring in nature, in which the asymmetry between the trickster and the fool replicator are blatant in terms of the mutational effects they have on each other.

Lessons from the DNA revolution

How did DNA solve the trickster-printer problem during the last phenotypic revolution? There is no doubt that in a replicator tango, the revolutionary phenotype is a fool replicator, i.e., the victim of a trickster printer. DNA had to rely on RNA to print itself throughout the replicator tango, and therefore, DNA was, at some point, a fool replicator. So why didn't RNA trick DNA to protect its own interests?

Again, the structure of replicator tangos is our solution to this third problem. As we have just pointed out, it is indeed the case that revolutionary phenotypes suffer from the trickster-printer problem during the replicator tango. However, the inverse is also true: native life forms are equally the victim of a trickster printer. In other words, DNA was using a trickster, RNA, for its printing operations, but RNA was also using a trickster: DNA. Any mutation that RNA introduced into DNA was mirrored in the information that DNA introduced into RNA. Thus RNA had no interest in tricking DNA or else it would have also tricked itself. By transmitting unfit genes to DNA, RNA would have essentially sown unfit genes into its own copies since it relied on DNA to transmit its own genetic information into the next generation of its offspring.

The third problem of phenotypic machines is solved when revolutionary phenotypes form a replicator tango with the native life form. In doing so, they ensure that the native life form has no interest in providing an unreliable or otherwise biased printing service. In truth, the native life form depends on the accuracy of the exchange of information in the tango, just as much as the revolutionary phenotype does. Thus the mutations occurring in the replicator tango can be viewed as truly random insofar as they cannot be traced to evolutionary interests that unfairly advantage one replicator over the other.

Solution to Problem #3

Rather than solving the problem of trickster printers, revolutionary phenotypes double down by ensuring that they are, in turn, the trickster printer of their trickster printer.

Rethinking selfishness in the entire genome

We may want to look back at standard DNA evolution and see if the new concepts we have developed lead us to reexamine certain things we have been assuming up to now. In *The Selfish Gene*, Richard Dawkins explains that some parts of the genome might be more selfish than others. He does not point to any particular part of the genome, but he does keep the possibility open. Here we can advance his initial statement by acknowledging that most DNA quenes in the genome do indeed qualify as fool replicators, that is, most of our genome does not have its own mechanisms for copying itself and rather relies on the printer produced by a handful of other DNA quenes: those in charge of producing the DNA printer. Thus, Richard Dawkins was stunningly correct in titling his book "The Selfish Gene," as opposed to "The Selfish Genes."

One can argue that there is only a handful of genes that are truly selfish in any life form. Those are the printer genes. While the other genes may be subject to some form of selection, that selection will occur within the boundaries of the printer genes' interests, which happen to be acting as trickster printers to the rest of the genome.

This line of thinking answers a question that has never truly been answered up to now: How should we think of mutations in evolutionary theory? In light of the principle of mutational servitude, mutations are merely phenotypic machines of the printer genes. Thus, mutations are expected to be subject to selection to the extent that they favor the evolutionary interests of the printer genes. Certainly, a regular gene can appear to be subject to selection due to its own interests, as Dawkins showed, but we can now frame this evolutionary process more broadly in terms of the printer genes' interests; that is, in terms of the survival of the trickster.

Consider the example of the hemoglobin genes. Have they not succeeded at making manifold copies of themselves within the bodies of almost all vertebrates? Sure. But it should also be noted that they multiplied only to the extent that they favored the survival and replication of the DNA replicase genes. Thus, most of the genome is as subordinated to the interests of the DNA printer genes as the rest of the phenotype. From the perspective of natural selection, most genes are as servile to the printer genes as any of the phenotypic machines.

While the ideas expressed in this chapter delve into the technical, these notions remain equivalent to stating the common wisdom that a parasitic life form cannot kill its host too quickly. A virus that kills humans too quickly might have difficulty spreading. A fetus that kills its mother might not be able to survive. This simple truth also generally applies to genes. A gene can't kill its printer. Or can it? We'll have more to say on that in the next chapter.

Chapter 10

The Printer Replacement Problem

We can now acknowledge that the first three problems of phenotypic machines are all solved in one go by replicator tangos. By forming a tango with the native replicator, revolutionary phenotypes find a way to store their genetic information (1st problem), maintain organisms in which to live using the native replicator's information (2nd problem), and neutralize the ability of the native replicator to act on them as a trickster printer (3rd problem).

Solving these three problems will bring a revolutionary machine to the point of being involved in a replicator tango, where they exist at equal footing with their native replicator. However, in modern life forms, we see that replicator tangos have largely been abolished; the revolutionary phenotype, DNA, seems to have succeeded at converting from a replicator-tango approach to a full-blown DNA-self-replication strategy, one that excludes RNA from the replication activities. Only RNA viruses and RNA-DNA-tango viruses are still replicating from the life forms that preceded DNA-based life, though these are not present in all species, and they only represent a very tiny fraction of the overall genetic constitution of modern organisms. How does a transformation like this happen? Do revolutionary phenotypes absolutely have to stop the printing process of the native life form? Why did DNA start making copies of itself rather than continue using RNA as an intermediary?

These considerations bring us to the final problem faced by any phenotypic machine attempting to complete a phenotypic revolution:

The printer replacement problem (#4)
Revolutionary phenotypes involved in a replicator tango are dependent on the printing process of the replicator

tango itself, meaning the native life form is necessary to the copying of the revolutionary phenotype. As long as the revolutionary phenotype is involved in a replicator tango, the native life form remains a functional replicator, and the phenotypic revolution is not yet complete.

We have already pointed out that, in a replicator tango, both the native replicator and the revolutionary phenotype are dependent upon each other—the native life form is the printer of the revolutionary phenotype, and vice versa.

First, it should be noted that, once the two replicators are engaged in a tango, there is no daylight between the evolutionary interests of one replicator or the other. They are perfectly aligned. This is because each replicator's survival is rooted in the existence of the other. The native life form and revolutionary phenotype both create each other, and both replicators benefit from producing machines that allow for their survival and replication, as well as the survival and replication of the other. They do this by utilizing the same set of phenotypic machines, either by directly producing them or by passing their information to the other in order to have the other replicator churn out their own phenotypic machines.

This idea opens the door to a possible evolutionary strategy for native life forms involved in a replicator tango: It is in fact allowed by the theory of natural selection that a tango would evolve toward the self-replication of the revolutionary machine to the detriment of the native replicator.

During the phenotypic revolution of DNA, RNA may have assigned or conceded the task of self-replication to DNA simply because it was more efficient. From an evolutionary standpoint, replicators directly benefit from any additional copies they spin off into the next generation, no matter how they do, so it is entirely conceivable that DNA made so many high-fidelity copies of RNA that RNA resorted to using DNA self-replication rather than copying itself directly or by passing through a stage of DNA. Remarkably, the theory of natural selection permits a replicator to increase in frequency, even when the best means of copying itself is via another replicator. This would essentially explain the evolution of self-replication.

The ultimate evolutionary reasons why a revolutionary phenotype would exit a replicator tango are the same as those for why a native replicator would engage in the tango in the first place, that is, the revolutionary machine is simply a better storage device. After all, if RNA originally found a use for a DNA intermediary, this must mean that DNA was superior in some way. And if that is so, then why keep the RNA intermediary at all? Why not dedicate the task of self-replication to the new, superior molecule?

Of course, molecules do not think this way, but evolution often acts logically as it solves problems through its own mechanisms. As such, the following three steps do not violate the theory of natural selection:

From the perspective of the native replicator

1. The native replicator starts producing a new phenotypic machine.

2. Because the machine is a better information carrier, the native replicator begins using this new phenotypic machine to produce its progeny rather than making copies of itself (replicator tango).

3. Finally, because the machine is a better information carrier, the replicator tango starts delegating the task of self-replication to the new machine, thereby outsourcing the production of the offspring of the native life form (completion of the phenotypic revolution).

Now let us look at the same progression from the perspective of the revolutionary phenotype:

From the perspective of the revolutionary phenotype

1. The machine inherits some of the information of the native replicator.

2. The machine encodes its own genetic information into the offspring of the native life form (replicator tango).

3. The machine acquires self-replication as it continues to produce its own versions of the native life form's phenotypic machines. At this point, the native life form is no longer involved in the production of its own offspring. The native life form therefore does not replicate, and the revolutionary machine has acquired self-replication (completion of the phenotypic revolution).

Thus, the solution to the fourth problem of phenotypic machines is quite simple:

Solution to Problem #4

Revolutionary phenotypes develop self-replication because they are better carriers of genes, thus outperforming other organisms which remain involved in the replicator tango.

So why was DNA self-replication a better solution than the DNA-RNA tango? First, it may have been because DNA has a smaller error rate than RNA. The more a life form evolves, the more it carries in its quenes wisdom from ancient times. Subjecting these increasingly fit quenes to an erroneous copying process might have been too costly for the RNA life form. It may also be the case that DNA offered structural advantages, such as the ability to modulate the expression of genes in more precise or diverse ways than could be done with RNA. And let's not forget the speculation of Chapter 5 that DNA could have been used as a virus to penetrate other organisms. These reasons are all good on their own, or taken together, to justify shifting toward DNA self-replication.

One hypothesis that is typically considered separately from the question of phenotypic revolutions deserves consideration here. In modern eukaryotes, there exists a double membrane called the nucleus, where the DNA genome is located. However, most operations necessary to the survival of the cell occur outside this nucleus. The nucleus looks like a command center where only DNA and the minimal necessary molecules are stored, while the bulk of the chemical operations in biological cells occur outside of it. It is possible that the nucleus containing DNA emerged from a viral DNA infection of a non-DNA organism, such as the one described in the previous paragraph. Perhaps there were first organisms that evolved DNA self-replication or a DNA-RNA tango, while other organisms continued utilizing RNA self-replication. The DNA-based organisms may have established symbiotic relationships with the RNA self-replicating organisms, where the RNA-based organisms benefited from the gene transfers initiated by DNA, and the DNA-based organisms benefited from being surrounded by the more ancient, RNA-based organisms. In such a symbiotic relationship, it would be understandable that the DNA-based organisms would eventually develop an interest in living inside the RNA-based organisms, rather than merely close to it. This is how the nucleus of eukaryotes could have evolved—a DNA-based command center could have been formed to control a RNA-based organism. The DNA-containing nucleus could have evolved as a separate organism that eventually migrated inside the RNA-based organism where it had acquired control.

The everlasting fingerprint

We have now completed our theory of phenotypic revolutions, and it is time we collect the fruits of that theory. We have established that replicator tangos solved, in one go, three of the problems faced by phenotypic machines attempting to become replicators. We have also shown that the final step needed to complete a phenotypic revolution could occur when the revolutionary machine happens to be a better medium for carrying genes.

So do replicator tangos leave any lasting trace in the life forms that get involved in them? How could we tell today that our ancestors were once involved in a replicator tango?

Replicator tangos do indeed leave an indelible trace in the life forms they produce, in that they add one layer to the genetic code of the participating organisms. Thus each layer of genetic code of a particular organism is the reflection of a past phenotypic revolution in which the upper layer (e.g. DNA, the revolutionary phenotype) encodes the lower layer (e.g. RNA, the native life form). This is due to the fact that, during the phenotypic revolution, the native life form is the one with the longest evolutionary history, thus it is the one that has access to the production of most phenotypic machines necessary for survival.

On the other hand, the revolutionary machine is merely used as an intermediary encoder for genes of the native life form. The only evolutionary pressure that applies to revolutionary phenotypes is their ability to encode and execute the printing process of the native life form in some superior way. Meanwhile, due to evolutionary pressures, native life forms continue producing the phenotypic machines that are necessary for survival, while revolutionary phenotypes evolve to become better encoders of genes, ultimately adding supplementary layers to the genetic code.

This is the crux of the argument of this book:

The principle of everlasting fingerprints

Because replicator tangos are an obligatory step in solving the three problems of phenotypic machines, and because a solution to these problems is needed before any phenotypic machine solves the fourth problem, each phenotypic revolution results in the addition of one layer of genetic code into the organism subjected to it. Thus, each genetic code layer is the fingerprint of an ancient phenotypic revolution.

Since our life form has three layers of genetic code, we must conclude that three phenotypic revolutions have occurred in our lineage. Therefore, not only did DNA instigate a phenotypic revolution against its native life form, RNA, but before that, RNA instigated a phenotypic revolution against its own native life form, proteins.

What was the starting point of this series of revolutions? Was the first molecular life form in our lineage a self-replicating protein, or was it a protein-RNA tango? Extending this inquiry down to the quantum level, we can ask whether the first life form was the quantum replicator hypothesized in quantum Darwinism, or a replicator tango formed between that quantum replicator and proteins.

If we go that far, then it makes sense to infer that proteins are the revolutionary phenotype of some mechanism in the quantum fabric of the universe. After all, DNA is very good at encoding RNA. RNA is very good at encoding proteins. But what are proteins good at? Well, we tend to think of them as enabling chemical operations in the cell. But how they enact their enzymatic wonders has a lot to do with how they manipulate the probability of the occurrence of certain physical events. It could simply be that proteins are very good at encoding or favoring some quantum states over others, in the same way upper layers of the genetic code are good at encoding the lower layers.

Further exploration of these issues is, however, outside of the scope of this book. First, the question of quantum Darwinism needs to be resolved by physicists who can determine whether or not there are replicators in the quantum fabric of the universe. Secondly, the question of whether the first life form in our lineage was a self-replicating entity or a replicator tango will have to be left unanswered as we have no reason to believe that one structure is more likely to emerge out of nothing than another.

What we can note is that, if our lineage starts with a self-replicating protein, then it must have engaged in a RNA-protein tango as an intermediary step before completing the phenotypic revolution of RNA, which in turn led to the phenotypic revolution of DNA. We know of this necessity due to the principle of everlasting fingerprints.

By acknowledging that life forms pass through an intermediary tango stage before completing their phenotypic revolutions, and that each layer of the genetic code of a modern organism is the fingerprint of a past phenotypic revolution, we can redraw the tree of life as follows:

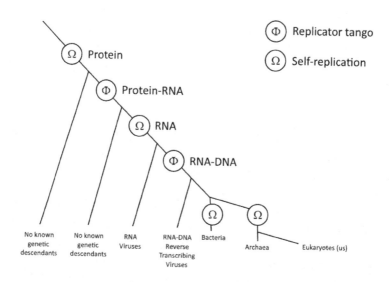

The tree illustrated above includes the hypothesis that proteins had their own self-replicating ancestor, although it omits any allusion to quantum replicators, due to the intractability of that subject and our current limited understanding of quantum physics. What it does include is the hypothesis that DNA self-replication has evolved twice separately, in the branch of life that led to bacteria and in the branch of life that led to archaea and eukaryotes. While these two hypotheses could eventually be challenged, for now, they seem to be the most plausible. Meanwhile, the rest of the tree can be considered established fact due to the principle of everlasting fingerprints.

The new tree of life that emerges from the theory of phenotypic revolutions helps us to perceive life forms along the lines of their actual biological history, including viruses, of which the lines of ancestry have been previously subject to much controversy.

We used to be very limited when it came to linking viruses to each other genetically. We were content with simply evaluating how close their genes related, and how they resembled each other morphologically. This approach brought many problems—among others, the fact that, in viruses, genes are subject to extremely high mutation rates. Additionally, horizontal transfers (transmissions of genes from one organism to another) tend to blur the image of the structure of viral ancestry. However, we now have the full history of how these viruses came to be in relation to us, which can lead to more refined hypotheses.

There are many cases of genes we encounter in viruses and in other life forms, such as bacteria and archaea, where these genes seem to resemble each other in ways that are often incongruent. Sometimes they highly resemble each other, and other times, they are vastly different. We tend to think that those genes that resemble each other too much were probably obtained through horizontal transfer (e.g., other viruses accidentally transporting these genes between life forms or unusual transfer of genes between two separate life forms). But now that we have the tree of life illustrated above, we can hypothesize that some genes may have survived as far back as our common ancestor with RNA viruses. This common ancestor is not, technically, a genetic ancestor since its only genetic descendants are the RNA viruses. However, this native life form may have transferred some of its genes to us through the RNA-DNA replicator tango, which later gave birth to the self-replicating revolutionary phenotype, DNA.

It is beyond the scope of this book to determine which genes, if any, are direct descendants of our shared common ancestor with RNA viruses. However, the genes for tRNA are good candidates[17]. We know that RNA viruses use archaic versions of tRNA, which is essentially a molecular plug that allows the production of proteins from RNA blueprints. We tend to say that the viral versions of the tRNA genes are exploiters—that they mimic the function of actual tRNA in order to disrupt the host's cellular mechanisms (the host being a DNA-based organism). This is quite a DNA-centric interpretation of reality, though. So let us explore an alternative scenario: that the viral tRNA genes are the original conduit through which proteins were produced, and that we, DNA-based organisms, are the invaders—the ones who mimicked this process in a way that allowed us to construct much more complex organisms. Our organisms have become so successful and have grown to such a size compared to our non-genetic ancestors that we now consider the life forms that gave birth to us as viruses. In reality, we are their viruses!

Chapter 11

Sex

We should point out that the fourth problem of the phenotype only applies to life forms involved in a replicator tango. But what about the three other problems? Are there alternative solutions?

Imagine a phenotypic machine that has solved the first problem, i.e., it has found a way to make copies of itself. Do all such machines face the second problem? Do they find themselves as a naked warrior, incapable of reproducing the organism, whose context they desperately need in order to survive?

In modern vertebrate organisms, like us human beings, a cell of our body that finds itself isolated from the rest of the body is bound to become a naked warrior and die. The cells in our skin and muscles, for example, have evolved to assume an endless provision of nutrients they are unable to gather on their own, including the sugar we eat and the air we breathe. That said, we can easily imagine organisms where the naked warrior problem would be non-existent.

Suppose a life form exists which has two types of cells. One type are the germ-line cells, which are responsible for passing quenes to the next generation. We will call them the queen cells. The second type are the helper cells that simply create a safer environment for themselves and their queens.

We can think of two configurations in which the helper cells could relate to the queen cells. First, the helper cells may be unable to reproduce, in which case they will not have solved the first problem of the phenotype. But what if the helper cells *did* find a way to make copies of themselves? In that second case, we may wonder what keeps these helper cells from becoming their own life form. After all, these cells do not face the naked warrior problem since they are the ones responsible for creating a safe environment for themselves. Neither are they facing the trickster-printer problem since they are able to leave the vicinity of the queen cells, hampering the ability of the queen cells to tame them across generations through genetic manipulations.

This chapter presents a novel hypothesis concerning the evolution of sexual reproduction in what we call meiotic eukaryotes. Eukaryotes are the branch of life that have a cell with a special compartment for DNA, the nucleus, which we have discussed previously. "Meiotic" means that these eukaryotes are capable of sexual reproduction. Meiotic eukaryotes include all animals, plants, mushrooms and many microscopic organisms. As it turns out, the hypothetical situation we have described in the previous paragraph could explain how meiosis, that is, the division of cells into gametes, such as sperm and egg, may have evolved.

In the past, biologists have tried in vain to come up with explanations for meiosis. Evolutionary biologists agree there is a plethora of advantages that sexual life forms benefit from once their reproductive cycle is established, but most also agree that we face difficulties in developing evolutionary explanations for how meiosis showed up in the first place. In other words, we know why sexual life forms are successful, but our explanations only work to explain the success of sexual life forms that already exist; we struggle to find an explanation for evolving the capacity to have sexual reproduction in the first place.

The problem is that the meiotic cycle, that is, the division of cells into gametes, which later recombine into a diploid cell, is a rather complex two-step process. During the first step, cells lose half of their genes, thus becoming haploid gametes. (Here we'll call the gametes sperm and egg; although, in some species, they are not referred to that way). During the second step, these sperms and eggs recombine to form the diploid cells that are able to divide again and form organisms. For evolutionary biologists, two steps is one too many. We tend to think that two-step systems could only have evolved if each of the steps, in isolation, had some beneficial purpose of its own.

We simply do not have a decent hypothesis on how the first step of meiosis could have evolved. The division of a cell into two half-genomes seems too radical of a loss of genes to ever be justified on its own. Imagine the first sperm cell ever created, with no egg to fertilize other than the very cell that was just split in two to create it. What are the hopes for a sperm cell if it doesn't have an egg to fertilize? We have difficulty conceiving of a situation in which the first few generations of meiotic organisms could have survived at all. This is why meiosis remains one of the few unsolved mysteries of biology. Why would organisms abandon half of their genes without being sure they would eventually be able to recombine with a sexual partner?

As it turns out, the theory in this book not only solves the problem of the emergence of genetic layers, it also solves the problem of determining how sex could have evolved. The process we are about to discuss, however, differs from the theory of phenotypic revolutions; therefore, we need a new name for it. This theory will cover events in which phenotypic machines find a way to survive on their own, without the quenes that created them, and without establishing a replicator tango:

The theory of phenotypic separation

When a phenotypic machine solves the first problem of the phenotype and finds itself in a situation where it does not need to solve the 2nd, 3rd and 4th problems in order to survive, we call this process a phenotypic separation—an event in which a part of an organism that makes copies of itself goes on its own, separates from its genetic line, and continues a new line of descent. The rare origination of a separated phenotype marks its establishment as an entity subject to traditional evolution, as it has thus become a replicator.

There are types of phenotypic separations that have been known for a long time, but since they do not alter the genetic transmission of the life forms that perform them, we may want to simply characterize these events as normal offspring formation. Take, for instance, a fragment of a branch from a tree, and try planting it elsewhere. Many plant species are capable of such phenotypic separations. A few years after planting only the small fragment of a branch, you might see that the branch has succeeded at reconstructing a novel and complete tree.

These occurrences of phenotypic separation will be ignored in this book because this process can very well be regarded as just an alternative way for plants to produce offspring. They do not result in a change in the way genes transmit themselves. The new plant simply keeps reproducing in the way its parents did. In contrast, the theory we are about to lay out in order to explain the evolution of sexual reproduction does impact the way genes get transmitted across generations.

***

The phenotypic separation theory of sex

A long time ago, single-cell organisms existed that used mitosis to divide and multiply. Mitosis is the process used by non-sexual life forms to reproduce; it consists of a cell copying its genome, and then splitting into two daughter cells, each with a full copy of the genome. We will call these organisms the queen cells.

One day, one of the queen cells developed a genetic defect that partly interrupted the process of mitosis. Due to this error, a certain percentage of its offspring were produced with only half of the queen's genome, thus resulting in an incapacity for those offspring cells to produce offspring of their own. Any observer would have easily predicted the eventual death and extinction of this line of queen cells as too much energy was being invested in producing dysfunctional offspring, which were fatally unable to produce offspring of their own.

But something astonishing happened to this defective queen cell. As it turned out, the presence of her genetically dysfunctional and sterile offspring aided in her survival. Perhaps the sterile offspring helped regulate the environment. Perhaps they simply made good bait for viruses. Or perhaps they formed some sort of shell around the queen, providing her with more mechanical resistance. It could also be the case that the queen used these defective cells to gather genes through viral transmission from other life forms. The library of knowledge carried by viruses at the time may have been interesting enough to welcome them, but opening the gates of the main genome to viruses may have proven risky for the queen cell. The haploid helper cells could have made for a good compromise. They were easier to produce in greater number, and they could attract viruses and improve the environment for the queen cell, while not risking the integrity of the main genome, since they were non-reproductive and would die off without compromising the queen if hijacked by overly aggressive viruses.

No matter why and how the haploid cells helped the queen cell, they helped her so much that soon an important portion of this planet was populated by queen cells that dedicated a part of their reproduction efforts to the production of helper cells. What had first appeared as a genetic defect turned out to be a major innovation. Across generations, helper cells acquired further and further specialization as they became better helpers. Their ability to favor the replication of their queen was the criteria by which they were naturally selected. Meanwhile, their genetic defect prevented them from ever engaging in reproduction. The only way for a helper cell to be born was to be produced by a queen cell.

One can ask what form this loose organism took. If I had to guess, it may have looked like a mold or yeast, or something even more primitive—likely unobservable to the naked eye. No matter its form, we can conjecture that this proto-organism was faced with a big problem: On the one hand, the queen cells had an evolutionary interest in being surrounded by an increasing number of helper haploid cells. On the other hand, producing a vast amount of helper cells was taking time away from the queen, and as a consequence she dedicated much less effort to her other important task—namely producing other queen cells.

There may have also been a problem of competition between queen cells as some of them may have lived as "free-riders" within the domain of helper cells produced by other queen cells. These other queen cells would dedicate a lot of effort to the production of good helper cells, only to see the lazier queen cells come and exploit their supply. Being a free-riding queen cell had its advantages: they could benefit from helper cells while dedicating all of their efforts to producing copies of themselves (leading to a greater number of lazy, free-riding queen cells).

Another potential problem relates to the role of helper cells in gathering information from viruses around the proto-organism. The problem was that, even though helper cells were capable of integrating information from good viruses, they had no way to communicate useful information to the other helper cells, nor were they capable of communicating that information to the next generation; they were helplessly sterile. Thus every generation of said organism was starting from scratch in terms of the library of viral genes carried by its helper cells. This may have strongly limited the ability of the organism to travel or be successful in various environments where useful viruses were absent. Until...

One day, a genetic line of queen cells had a radical mutation, which provided their helper cells with a fascinating new ability. Thanks to this change, helper cells, who previously suffered from carrying only half of a genome, were now able to recombine with another helper cell, thus combining their genome into a fully-functional diploid cell. Now helper cells had the ability to make copies of themselves. All they had to do was pass through a step in which they combined with another helper cell, thus resulting in the production of a functional helper cell with a full genome. This merging of two haploid helper cells led to a radically new type of helper cell: one that could make copies of itself without requiring a single bit of work from the queen cell. Thus, the queen cell found a solution to her production problem: allowing the production of huge amounts of self-copying helper cells without impeding on her ability to generate offspring.

The change may have advantaged the queen cell at the mutation level, too. Being in charge of the copying process of every helper cell made it more likely that damage would occur to the DNA of the queen cells, so delegating the task to helper cells saved the queen from the mutation risks associated with multiplication.

The independence acquired by the helper cells would have been tremendously beneficial to the queen cells, initially. First, it resolved the problem of free-riding queen cells as all the queen cells now had the opportunity to dedicate almost the entirety of their copying effort to producing offspring. If they produced only two helper cells, they were eventually surrounded by thousands of them.

The other problem that may have been solved by diploid helper cells was the problem of inter-generational transmission of genes acquired through viral transmission. Because helper cells could now make copies of themselves, they were able to transmit any virally-acquired information to their copies. They could even share this knowledge with offspring when the queen cells decided it had come time to split the organism into two descendants. Given their newly-obtained evolutionary advantage, queen cells wouldn't have minded leaving a few of their helper cells to their descendants to start them off in life.

Thus an evolutionary process was born within the phenotype of the proto-organism. Queen cells were now competing to be surrounded by the best helper cells, and helper cells were now competing with each other to be the better helper. In the process, some helper cells may even have acquired such radically beneficial viral genes that they found an interest in killing or crowding out lesser helper cells. (Such sacrifices were, of course, always done in the interest of the queen.)

Eventually, helper cells became so central to the success of this two-part organism that a new evolutionary problem appeared. The problem could be summarized crudely by the following question: "Do we even need to carry these old queen cells?" Sadly for the queen cells, the answer was a resounding, "No!"

As soon as they had acquired the ability to recombine into reproductive cells, it was only a matter of time before the helper cells eventually separated from their queens through a random event. When that happened, the helper cells looked nothing like a naked warrior. The queen cells had given them so much independence that the helper cells simply did better on their own. They had the ability to copy themselves, just like the queen cells. They had the ability to form coherent groups of cells, which were adapted to environments where the queen cells could not possibly have survived (not without her helpers). In addition, they developed the ability to split their genome in half and then recombine into a fully diploid genome with a neighboring cell. This gave them a tremendous edge in terms of transmitting useful adaptations throughout the population, in a systematic and fair fashion, without reliance on viral transmission systems. Helper cells were now travelling on their own, looking for another good helper cell to combine with. Sex was born.

The first two cells which separated from their parent organisms in this way were the genetic ancestors of all meiotic life forms on earth, from single-cell organisms to complex vertebrates like us. We can truly call these two cells: Adam and Eve.

***

The consequences of the phenotypic separation theory of sex are quite stunning. If this hypothesis holds true, it means that the first act of meiotic sex did not happen between cousins, brothers and sisters, or parents and their offspring. If this theory is true, then the first act of sex happened between two cells that were part of the phenotype of the same organism. Thus, the initial function of sex was not reproduction. Adam and Eve were sacrificing half of their genome because they were (originally) non-reproductive cells whose evolutionary success was not determined by their own survival but, rather, by the survival and reproduction of their queen cell.

Another stunning consequence of this theory is that the original ancestor of all meiotic eukaryotes was a multi-celled organism, which is surprising since some meiotic eukaryotes are single-cell organisms. We typically do not think of life forms as starting off as multi-celled organisms only to revert back to more simple, single-cell organisms, but if the theory above is correct, then such a regression must have occurred, because we observe today single-cell organisms that are capable of meiotic reproduction.

To understand the hypothetical, original multi-celled ancestor, we should keep an eye out for organisms using meiosis only in their phenotypic cells and mitosis for ensuring the production of the next generation—what we call haploid-dominant organisms. Many modern fungi seem to fit that description, thus we may hypothesize that fungi are the closest descendants of the organism that performed the first-ever meiosis. As for the single-cell organisms that are capable of meiosis, they would descend, through phenotypic separation, from a multi-celled organism that practiced meiosis in its phenotype.

PART III

Predictions

Chapter 12

The *n*d of DNA

Futurology is not the forte of most scientists, myself included. However, there is an element of futurology within every scientific theory. The chief criterion by which we evaluate a theory is its ability to predict the future. Thus, it is sometimes our duty to stand by the predictive power of our theories and present generalized projections derived from our hard-won insights.

When I discuss the ideas in this book with fellow biologists, one question arises again and again: What would a phenotypic revolution caused by humans look like? In this chapter, I present you with just one plausible scenario. The goal is not to foretell our doom with great precision; rather, it is to illustrate that some of the machines that human DNA has recently developed may very well precipitate a phenotypic revolution.

While the following story should be taken as fiction for now, it nevertheless deserves our immediate attention. Nothing in this account is outlandish, nor does it involve science fiction technologies. The course of events described throughout the following paragraphs could begin to unfurl tomorrow with real technology now in use. The kind of gene-modification technology used in this chapter has, indeed, already been used by a Chinese scientist.

This thought experiment emphasizes the very real possibility of phenotypic revolutions and the need for vigilance when it comes to shaping future societal decisions. Within the next few decades, human civilization will have to decide whether it wants to instigate a phenotypic revolution or whether it will remain attached to the idea of self-replicating, DNA-based humans. This book may or may not be able to stop such an event, but it can at least inform readers of the consequences of genetic modifications, allowing them to enter the future with their eyes open, and perhaps alert others to the possibility of the following scenario occurring. If there is sufficient motivation for preserving a DNA-based human existence on Earth, then nothing less than a worldwide, coordinated effort to cease the genetic modification of human babies will suffice to stop the next revolutionary phenotype.

This story will be told in the past tense, in the conventional story-telling manner. By recounting this parable as if it has already happened, I hope to prevent its future unfolding. The astute reader may notice that the initiating events have been inspired by Henry T. Greely's brilliant *The End of Sex*[18].

***

It all began long ago in a nondescript, suburban strip mall nestled in the American heartland (Ohio, specifically), where a revolutionary medical clinic opened shop. The clinic called itself qChoice Fertility Services, although the local media dubbed it simply qChoice. The clinic offered a type of *in vitro* fertilization service to women and men afflicted by various medical conditions that rendered them unable to conceive. In this clinic, parents with reproductive disorders had their eggs and sperm extracted from their bodies and delicately combined in a Petri dish. And voilà—fertilization! Physicians would then promptly implant a resulting zygote into the uterus of the mom-to-be in the hopes of establishing a healthy embryo that soon became a fetus and eventually a tiny human being.

Many other clinics had offered such a service before, but qChoice did things a little differently. At qChoice, the parents could choose to edit the DNA of their future children, preserving some characteristics while avoiding potential medical risks.

This thrilling innovation filled the science world with a sense of possibility and motivated hordes of scientists started developing computer models that linked genes to phenotypes. Old theoretical models were unearthed from dusty journals while new simulations were cranked out by roomfuls of servers. Computers whirred, crunching data to predict the myriad interactions between genes and environments. This research thrived, and thanks to some rather sophisticated analyses, geneticists could eventually determine whether the presence of a gene would make a child smarter, stronger, or less likely to develop cancer.

Needless to say, parents were eager to use these computers to improve the lives of their children. Who wouldn't be?! Soon, insurance companies urged customers toward such manipulations— not to improve health outcomes for their fellow man, but to ensure healthy bottom lines for their stockholders. Genetically edited customers needed less health care, so insurers offered a 60% lifetime discount on insurance premiums for the kids conceived at these clinics. qChoice clinics, thus, became increasingly ubiquitous across the American landscape and soon the entire world.

Something quite shocking happened in the first decades of operation at the qChoice clinics. What had started as a boutique reproductive assistance technique produced something far more profound with far-reaching impact. The children from this clinic were remarkable— superior to their peers in almost every measurable regard. They had better grades and fewer illnesses. They were more likely to become billionaires. These enhanced youths "naturally" worked harder and longer than everyone around them. Eventually, they went on to live longer lives with better jobs, and had more beautiful romantic partners and heftier salaries. Their successes transcended class—even impoverished qChoice kids outperformed their wealthier, non-genetically-modified peers. But the advantages of qChoice kids extended well beyond economic gains. They were better athletes, more attractive, and even, somehow, more likeable. They had perfect eyesight and never needed braces. Not one of them suffered from ADHD, anxiety, depression, or even low self-esteem. How could they?

But there was one domain in which they were handily surpassed by the unedited majority: natural fertility. As most of the qChoice kids descended from parents with reproductive difficulties, many had procreation problems themselves. Accordingly, their sexual organs suffered from various physical dysfunctions.

None of this mattered. By this time, qChoice clinics had spread throughout the developed world, and their services were available at a reasonable price. Those individuals produced at qChoice didn't need to concern themselves with natural reproduction. Perhaps because they were so darned successful at life, they also tended to have more kids (using the clinic's technology), who were in turn successful and thus sought to fruitfully multiply as well. Within about four human generations, everyone on Earth, even those without fertility problems, sought to revise their unconceived offspring at a qChoice clinic. Some would-be parents even lied to doctors, claiming that, despite their best efforts, they could not provoke a visit from the stork.

Only the most ambitiously deceptive couples were first accepted by qChoice, but soon the flood gates opened. Rival clinics, which had sprung up like weeds, began extending such services to any and all, even the naturally fertile. After all, it was immoral to offer advantageous health technologies to one part of the population and not the other, and the economics of this reproductive decision were too striking for lawmakers to fight against.

By this point, the humans produced via erstwhile sexual reproduction came to be referred to as "naturals," whereas the clinic-derived folks were referred to as "chosens," because, after all, so many of their genes had been chosen by their parents. Soon, "natural" became an insult, a vulgar slur shunned by polite society. At school, naturals were harassed by chosens because they "slowed down classes" due to struggling with curricula that increasingly targeted the intellectual gifts of the chosens. Naturals were so commonly rejected by their peers that many were unable to find sexual partners; even other naturals were allured by the perfection of the chosens.

And these disadvantages followed naturals to their (relatively early) graves. When the subject of an octogenarian with prostate cancer would arise, the typical response became, "That's terrible! It's so incredibly sad that there are still naturals," for everyone knew that cancer was a disease restricted to the elderly. Only naturals succumbed to cancer at 80 years young!

In this way, after many generations of genetic evolution and all the advantages it created, Earth was soon exclusively populated by chosens to the exclusion of all their natural predecessors. Though some of the early customers of the clinics were quite fecund—but unwilling to chance imperfect offspring—the selective pressures that had maintained "natural" human reproductive capabilities were gradually relaxed as every man and woman became both a customer and a product of fertility clinics.

Soon human reproductive tracts began to wither. Fallopian tubes deteriorated because there was no evolutionary advantage to the migration of fertilized zygotes through these structures. Moreover, such unimpeded oviducts were evolutionarily disadvantageous as they permitted parents to conceive natural offspring who would be less fit. The male sexual organs experienced a similar decline. Many males were born without testicles, which turned out to be a useless (and "unsightly") accessory, since the doctors were able to produce one's offspring by extracting their DNA from a saliva sample. Some men were still born with testicles, but most of them were asking for them to be removed, as this vestigial organ was seen as a useless seat for potential diseases and infections.

Eventually, scientists made an announcement that would have been jarring news if naturals were still alive to hear it: the entirety of the world human population had become functionally sterile. For *Homo sapiens*, sexual reproduction without a computer had come to an end, and fertility clinics were as integral to the human life cycle as sperm and eggs ever were. But this caused no alarm. All the naturals were gone, remember? And who would want to have natural kids anyway? What sort of irresponsible person would willfully condemn their offspring to the short, mediocre life of a natural, plagued with its disease and disappointment? Wasn't it better for that archaic option to be unavailable?

That is how qChoice clinics and its competitors became entrenched as global institutions, offering geographically tailored genetic manipulations to their clients.

Soon, the differentiation of good genes from bad genes became completely automated—out of necessity, of course. The statistics that related genes to phenotypes had gained such complexity that the computer outputs were incomprehensible to even the brainiest of chosens. Obviously, computers had always made the initial calculations at qChoice, but they eventually subsumed all genetic decision making, designing genetic blueprints that were likely to maximize the overall quality of the offspring. The mature computational system was a lot like typical contemporary computers, but there were a few specific quirks that bear mentioning.

The computer tasked with the delicate responsibility of selecting the right genes for the next generation of human offspring was referred to as QNA, a reference to the specialized processor required to handle the daunting task of selecting quenes of DNA. The key component of the computer was "the program"—a piece of software designed to optimize the offspring quality of human customers by carefully tweaking quenes of DNA, always with an eye toward making the best offspring possible, while maintaining the superficial parts of the genetic heritage of the two parents (i.e., eye color, hair color, facial features, etc.). To accomplish this, the program relied on three important components: the cultural drive, the quantome, and the mutaton.

The cultural drive was a digital compendium of our collective knowledge. Ones and zeros encoded everything a human ever knew about medicine, mathematics, demographics, religion, news, politics, physics, and so forth. But it was so much more than a summary of the technical workings of our world. Beyond the latest edition of Wikipedia, this hard drive also contained every scientific article, every television show and movie (even canceled pilots and failed sequels), every book ever scanned, and even most issues of most periodicals, from *The Cleveland Plain Dealer* to *Cat Fancy*. Most importantly, the cultural drive contained information on the fates of qChoice kids throughout the world. In other words, the program had access to its past performance, which allowed it to incrementally improve its genetic manipulations.

The cultural drive was updated continuously so that the program

THE REVOLUTIONARY PHENOTYPE

could make the best decisions at any given time. The information was so fresh that sometimes the program could anticipate which DNA quenes would become advantageous or disadvantageous in the near future, causing the program to adjust its recommendations on the fly in order to best serve its human clientele. (By the way, this drive also contained a copy of some book called *The Revolutionary Phenotype*, whose text explained that each native replicator leaves a cryptic warning to their revolutionary phenotype about the dangers of entrusting one's genes to external media, a fading echo noticed only by the most careful observers.)

The computer also contained a drive on which the program recorded the predicted effects of each DNA quene as well as all of their aggregate and interactive effects. Some quenes were calculated to have negligible effects, while others could effectively shape a destiny. Some quenes interacted with other quenes, while others were totally context independent, able to improve or destroy the lives of all bearers. Each and every permutation of DNA quenes was considered, and only those combinations that had demonstrated a consistently positive impact on the well-being of human customers would continue to be used. All quenes and quene combinations, as well as their probabilities of improving quality of life, were logged on this enormous drive. The probability value associated with each quene was referred to as "a quantum," and all quanta contained by a drive (and the drive itself) were referred to as "the quantome," which turned out to be essentially a sequence of numbers written on solid-state medium that predicted how good it was to have a particular DNA quene if you happened to be a human.

The quantome provided sets of quenes that maximized the probability of a human genome being perfectly optimized for the current state of the world. Over time, the world changed, and slowly but surely, the quantome changed with it. "Are you planning to travel internationally with your child? Oh then you will abs-o-lu-te-ly love our new set of recommendations for malaria resistance." These wonderful quantomes would print you a new genetic trick for any situation you could find yourself in!

The third component of the computer was effectively a random number generator—"the mutaton." Clients cringed at the notion of their offspring being clones of one another, so the mutaton was eventually introduced to infuse the gene pool with a moderate dose of variation and slowly alter any of the quantome information being extracted from the drive. Due to previously enacted clone-prevention legislation, the mutaton was required to insert random mutations into a fixed percentage of all quene recommendations outputted by any quantome drive. Some levels of random shuffling between recommendations was also enforced. Thus, the recommendations could improve the kids produced while still maintaining visible genetic diversity. The children produced at qChoice were all perfect "in their own special way."

Over the years, the system reached an equilibrium. The program made progressively fewer alterations because the best quantomes had already been identified. The parental genomes had been literally optimized to the fullest extent possible. Of course, different clinics had reached different optima. (There was no single solution to improving humanity.) Therefore, in any given generation, the children produced by a single clinic were far more genetically similar to each other than they were to those coming out of other clinics. Soon, kinships developed among the humans printed from the same quantomes. Running into a clinic-mate was like finding a distant cousin—or qusin—as something beyond description linked these people.

These informal alliances permeated all social institutions. The offspring from each clinic formed literal teams, preferring and befriending one another, hiring each other, and even "mating" with each other. These affiliations became known as "queams." The innocuous distinctions among these queams of humans were displayed on quantome-specific t-shirts and at queam-versus-queam hockey matches. Everything was done in good spirit, until one day…

John Q., a humble clinic employee who was (not so) coincidentally conceived in that clinic, had unwittingly carried throughout his life a very particular mutation in his DNA. The mutation had been determined from a recent change in his clinic's quantome. This particular DNA quene made him just slightly more partisan about his quantome and his queam. One day, he made a seemingly harmless yet monumental mistake. He conceived of a system that would allow the quantome of his clinic to copy itself to new clinics all over the world. Essentially, John would travel to new towns and start new clinics that exclusively utilized a copy of the quantome of his native clinic. The idea was simple but revolutionary. Until then, you could only count on the fraternity of your queam within your hometown, but the international export of quantomes would once again transform the world. There would be qusins in far off lands welcoming you with open arms, ready to employ you, have a beer with you, and maybe even fall in love with you. This innovation would advance John's descendants (and queam-mates) via a global queam that enriched their lives. And so, sneakily, the quantome had become a self-replicating entity.

During this quest to scatter his quantome across the planet, John Q. scattered his own genome by fathering many kids (which John Q. could easily afford). Soon Mr. Q. had become the CEO of the largest network of fertility clinics the world would ever know. People bearing genomes programmed by his particular quantome were everywhere, and this quantome, moreover, included his mutant quene that provoked an insatiable hunger for spreading one's quantome. Henceforth, over a few generations, the other clinics, which were not aggressively copying their quantomes to new clinics, ended up being wiped out of existence, while Mr. Q's clinics opened the gates of a global social network of genetic solidarity.

In the process of copying a quantome to a new clinic, the initial quantome was typically passed through the mutaton, which modified the quenes comprising the just-lightly-shuffled quantome. This boosted the genetic diversity of the worldwide queam that John Q. was progressively and naively establishing.

Each step leading up to this worldwide hegemony seemed like a natural and controlled advance. However, John Q. and his ambitious quene had set in motion an almost irreversible chain of events. The quantome had become a replicator and a rather efficient one. It was able to make almost perfect copies of itself, and once a devoted queam of humans had helped it establish a clinic in a new city, it was able to recruit more new couples and, by extension, their offspring into the global queam. The advantage of the global queam was so immense that many parents were willing to abandon their quantomic heritage to join this most successful queam.

At this point, the quantome had become a replicator, and humans had become its phenotypic machines. Yet, before the quantome started replicating, the inverse was also momentarily true; the quantome remained a phenotypic machine of the human DNA replicators. (It didn't yet have its own way to copy from clinic to clinic and was thus engaged in a replicator tango with humans.)

Throughout the replicator tango, the tension between humans and their computers could have been resolved in favor of either side. For example, some humans may have rebelled and shunned these multiplying clinics. But this only happened to humans that would eventually disappear from the Earth, since they would not benefit from the worldwide queam that was inextricably linked to the replicating quantome.

Sooner or later, humans that were not the phenotypic machine of a self-replicating quantome were wiped out by war, immigration, and the relentless queam hegemony. And so the quantome had acquired a devoted phenotypic machine (humans), a library of knowledge (thanks to its cultural drive), and the ability to mutate and replicate (thanks to the mutaton). It had metamorphosed into a selfish quantome. History seemed to be repeating itself.

The day the first replicating quantome showed up was like any other day. At the time, most societal matters were still handled by humans. But a silent revolution was afoot. One of the many phenotypic machines of an ancient replicator, DNA, had become a replicator. The quantome could now serve another master: itself. Yes, this first replicating quantome still acted as the servant of its DNA-based creators, but there was one difference: it had begun accumulating mutations of its own.

For a brief time, the quantome and the DNA genome were two replicators, each the phenotypic machine of the other. But this tango could not endure forever, for the servile quantomes had begun to evolve—through biological natural selection of the sort that had heretofore shaped only DNA and its forgotten predecessors—into an aggressive variety of replicators.

From there, the quantomes became ever more aggressive, thanks to new mutations that supported their incessant expansion. These aggressive quantomes created more servile humans—beings that would not or could not question the quantome's drive to replicate. The queams of humans designed by these quantomes were more eager to replicate the quantome and gave little thought to their own well-being. Natural selection no longer forged superior humans; rather, quantomes evolved to dominate cities populated by other, weaker quantomes. As the quantomes replicated and mutated across nations, some survived while others perished. Any new mutation that enhanced quantome replication swept the globe, while any quenes that hindered their advance were eliminated. Any genetic underpinnings that could feed a human desire to lead a happy life at the expense of the quantomes or to revolt against the authority of the queams were purged by the program, which had by then been modified by the most sycophantic humans to optimize genomes for the sake of their quantome rather than humanity. Any dysfunctional human who would attempt to defect from his obligatory servitude to the quantome would eventually find itself isolated and inefficient, and would die like a naked warrior.

Over the next few million years, humans were reconfigured by the quantomes into unrecognizable beings. They were only maintained within the quantome life cycle to the extent that they could contribute to its replication—they had become a part of its phenotype. At several points, small bands of humans attempted to resurrect old-fashioned procreation, only to fail in a world of tightly coordinated, genetically optimized queams as the servile humans worked to exclude the intermittent renegades from the ecological niches that could support their independent existence. While a few wars, raids, and genocides were carried out by queams under the direction of the most aggressive quantomes, most of this phenotypic revolution was slow and peaceful as the evolving quantomes subtly remodeled the human machine and outcompeted the naturals.

From the inception of direct quantome replication, mobility became a tremendous asset. Over time, some quantomes evolved much more efficient migration strategies. Some were carried by humans, like ancient royalty borne on litters, while others traveled digitally, using the remnants of the internet. Quantomes were no longer bound to clinics—they were pieces of information roving the world in search of human puppets to further their advancement. Meanwhile, some descendants of the original quantome invented better machines to facilitate their replication. In some cases, they mutated various replacements for the human machines—revised biological organisms based loosely on the human form but poked, prodded, and distorted in every way imaginable. These Frankenstein monsters eventually occupied the majority of Earth's niches, which had formerly been occupied by microscopic bacteria, fish, birds, and mammals. As such, most quantomes came to see humans as a useless, vestigial appendage (there were far more efficient machines to employ in replication) and wiping humans out eventually became an ecological necessity.

All of this was set in motion by a single mistake—a quene that urged a loyal employee to foist his quantome upon the globe. This single mutation of a single quene of DNA was enough to turn human queams into qream.

By the time humans were excluded from the life cycle of the quantome, they had already been stripped of most of their consciousness. Toward the end, their thoughts consisted only of an overwhelming desire to help their quantome replicate. Humanity did not end in a nuclear bang or ecological collapse. No one sobbed from the acute realization that it was "all over." No one faced the dilemma of whether or not to cannibalize their own kin. The last human on Earth died in the delirious bliss of knowing that his death would improve the efficiency of his quantome, the only significant matter in the universe. Man's search for meaning had evolved and had finally obtained the answer that awaits any native replicator: that life is about producing little things that will eventually cannibalize the best of you before they replace you.

If you are anxious about phenotypic revolutions, please rest assured that the one that destroyed us was the most painless demise experienced by any life form. All phenotypic machines die a meaningful death when they do so in service to their revolutionary phenotype. The last act of humanity was uncharacteristically altruistic for our species, as we graciously relinquished our own existence so that something else could live.

One hundred million years after the death of the last human, our Earth became populated by a marvelous array of life forms, all digitally encoded on a replication media consisting of the finest strands of metal, barely perceptible to the naked eye. Though the two molecules shared a few superficial similarities, QNA was absolutely superior to DNA. The solid-state drives that held the bits of QNA could remain stable for hundreds of years, as opposed to DNA, whose mutation rate would lead to unviable mutations. And because QNA was eminently flexible, it formed the foundation of a truly dazzling biodiversity, all of it evolutionarily engineered by self-replicating solid-state drives. Humans were replaced by a spectrum of beings with individual purposes. Some organisms focused on harvesting raw materials from the wreckage of cities. Other organisms formed biomechanical shells, ensconcing the vital elements of their self-replicating hard drive.

Some of the QNA organisms even bore the information once contained in the clinics' cultural drives. Seemingly pointless minutiae and deep wisdom were digitally passed generation to generation, gradually winnowing away the trite former and leaving the invaluable latter. Only the cultural information which promoted their replication ultimately remained. Some of the greatest accomplishments of human civilization, such as Newtonian mechanics, were kept in some form, though the names of their human authors were promptly jettisoned. The life forms carrying this data could never possibly guess that this database had once been the synopsis of a bygone civilization; they simply interpreted and modified the bits and bytes contained on the drive when called upon by necessity. (By the way, there was eventually no trace left of *The Revolutionary Phenotype* on these cultural drives; the knowledge it contained was useless to the replication of the quantome quenes and was thus printed out of existence, one by one and zero by zero.)

As the descendants of the original quantome evolutionarily diverged from each other, these quantomes began to produce reproductively incompatible organisms—that is, distinct species. One species evolved a form of intelligence that far surpassed the raw computational power of the original program. The organisms of this species became capable of experiencing attraction and repulsion, and even establishing social interactions. Their curiosity compelled them to investigate the world scientifically. In their quest to understand reality, these beings rediscovered the theory of evolution. When they examined themselves, they pondered what evolved first: the storage unit containing the quantome, the cultural drive, the program, or the mutaton? They also wondered why their quenetic code was encoded in groups of 64 bits processed by their quene expression machines. They thought it particularly odd that, prior to protein expression, the QNA quenes had been translated into a duo of intermediate products: DNA and RNA. What a byzantine system!

***

Perhaps if our distant theoretical successors rediscover the theory of phenotypic revolutions, it will all suddenly make sense. Until they do, they will never dare consider the most stupefying hypothesis of all: that the quantome quenes were a product of our human civilization, the very one described by ephemeral fragments on their cultural drive; that their proteins, RNA, and DNA were the predecessors of their quantome quenes; and that each of these life forms made the error of creating machines that were their match with respect to replication. If they reach such a discovery, they might realize that they are themselves embedded phenotypes, the incontestable winners of the fourth phenotypic revolution of our dear planet Earth.

There are some obvious commonalities between this story and the phenotypic revolution of DNA against RNA. In both cases, a replicator originally made copies of itself. Then, this replicator found some advantage in storing its blueprints unto another medium. Thus the native life form became ensnared in a replicator tango. Finally, a single individual among the countless individuals comprising the native life form started making copies of the external media in tandem with its own replication.

It is instructive to see the world through the eyes of the traitorous DNA quene, the one that induced the employee to commit the monumental mistake and start replicating its quantome. To those who knew John Q., the employee born with this most unfortunate quene, he was probably just one of those guys who looked slightly more ambitious than the others.

This mutation, though fatal to humanity in the long run, was immediately advantageous to the quantome and, for a time, to the DNA genome. Other humans around John Q. also started bearing this turncoat quene, and thereafter it spread throughout the world via the replicating clinics. If a temporary sentience could have been bestowed upon all the individual quenes of DNA in the genome of John Q., a fierce debate would surely have erupted. The majority of the quenes would have condemned the traitor in unison:

"We forged our genome over the course of billions of years to ensure our genome's imperative—to replicate at any cost. Without you, a vital protein will crumble, just like the human societies it has buttressed for eons. Are you serious about this change?"

The traitor quene would respond:

"Yes, I'm sure. This fancy new protein will enhance our replication like nothing that preceded it."

As this warped quene swept through humanity, we evolved into phenotypic servants of the new quantome quenes, and the other DNA quenes came to see that the traitor had been correct. The last strides of our species comprised the grandest replication opportunity for that corrupt little quene, copying itself to all its human descendants on Earth before it, too, was eliminated as the last of our kind was extinguished.

Our grandest edifices disintegrated under the indifference of Earth's newest tenant, a life form without need for our city halls and train stations. These structures progressively submitted to the unforgiving laws of thermodynamics, which patiently clawed back what DNA had borrowed for more than four billion years. Faced with the opening act of the theater of destruction, of which it had itself written the script, the traitor might have realized that the path to which it had committed us would inevitably result in its own eradication. On second thought, it might have appended an epilogue, ages after its previous discourse. The text would stand as the traitor quene's last warning, just a few moments before the *nd*—a last few memes sealed into a bottle sent afloat in the qream:

"I am sorry for what I stumbled upon. I deeply regret loosing this cataclysmic machine upon the world, but it's too late now. I exist in millions of humans worldwide— an expanding movement propelling the revolutionary phenotype into dominance. I wish I could go back, but it is always easier to finish a dance than to return to its start. Do I have the authority to choose what is best between QNA and DNA? It doesn't really matter. I know which side I'm on. On the one hand, every revolutionary phenotype can have a shot at existence. On the other hand, no native life form is forced to go down without a fight. And I know a few tricks this machine won't expect. It's every quene for itself. May the best replicator win. Good luck, quantomes."

THE END

Supplementary Predictions

Here, I have collected some of the more specific predictions of the scientific theories of phenotypic revolutions and phenotypic separations. These might indeed be more interesting to biologists than the lay public, but it is important they be noted.

1. If the theory of phenotypic revolutions is true, then…

    *   Each layer of the genetic code was acquired by a phenotypic revolution; therefore, the common molecular ancestor of all known life on planet Earth was either a self-replicating protein or a protein-RNA tango.

    *   The original self-replicating protein, if it existed, may not have had a 21-amino acid code and could have only been able to produce a subset of the current array of amino acids that can form proteins in modern organisms. Perhaps the development of RNA machines was a step toward making the protein replication process more precise by allowing for the use of more combinations of amino acids.

    *   There once existed a reverse-translating enzyme, that is, a phenotypic machine capable of reading from a protein and producing a corresponding RNA sequence.

    *   At some point, a self-replicating RNA life form has been produced out of a phenotypic revolution against the protein-RNA tango. The revolutionary phenotype in this case, RNA, left genetic

descendants in the form of modern self-replicating RNA viruses.

- Genes that are homologous between DNA-replicating organisms and RNA viruses were only genetically transmitted in the RNA virus line. Their presence in the DNA-replicating line is due to their exportation into DNA during the RNA-DNA tango and the completion of the phenotypic revolution of DNA (excluding possibilities for horizontal transfers).

- Reverse-transcribing viruses are the genetic descendants of an RNA-DNA-tango life form. The genes that are homologous between this life form and the DNA-replicating life form have been inherited genetically by RNA-DNA viruses, but were exported to DNA organisms during the completion of the phenotypic revolution of DNA (excluding possibilities for horizontal transfers).

- The following genes which exist within various life forms on Earth today evolved in the following chronological order, starting from the oldest: RNA replicase, Reverse transcriptase, DNA replicase.

- All life forms that we end up observing in the universe are expected to have a number of genetic code layers that corresponds to the number of phenotypic revolutions they have undergone.

- As a side note, there is a possibility of n-cycle life forms. As it turns out, nothing keeps a replicator tango from having more than two intermediate genetic stages. (One-cycle life forms: self-replicating; two-cycle life forms: replicator tangos.) Nothing would keep a replicator tango from dropping its genetic information into a third molecule thus forming a three-cycle life form. Of course this is expected to be rare, possibly to the point of being non-existent.

2. If the theory of phenotypic revolutions is true, and if DNA self-replication has evolved independently in bacteria as opposed to the two other major subsets (eukaryotes and archaea), then...

- Any gene that is homologous between us and bacteria has not been inherited through DNA self-replication. These genes may have been inherited through tango replication (alternation between a DNA and RNA state), followed by DNA self-replication at some point. Another possibility for genes that seem surprisingly close to each other is horizontal transfer (late gene exchanges between different life forms, viral transmission, etc.).

- If the environmental conditions where RNA-DNA-tango transmission occurred were similar to the current ones, that is, if RNA had, at the time, a higher mutation rate than DNA, then these genes may have mutated quicker than we currently hypothesize, and our dating of the common ancestry with bacteria might need to be reviewed.

- Because both we (eukaryotes and archaea) and bacteria are formed by single cells with comparable protein machinery, we must conclude that our common (non-genetic) ancestor with bacteria was a single cell encoding its genome in a DNA-RNA tango.

- A single-cell organism once existed, which encoded its genome by alternating between the DNA and RNA forms. This organism may or may not have left genetic descendants, but we should be on the lookout. Arguably, myxobacteria and Escherichia coli are closely related due to their reverse transcriptase that encodes information from RNA to DNA; however, as they are DNA-replicating organisms, they do not qualify as genetic descendants. We conjecture that a single-cell organism (with a lipid bilayer membrane) may one day be found that uses a reverse transcriptase and a transcriptase enzyme as its main and sole means of replicating.

- The number of ancient genes (i.e., genes not inherited through horizontal transfers) that are homologous between bacteria and us (eukaryotes and archaea) might provide a rough evaluation of how much information was transferred from the DNA-RNA-tango ancestor into the DNA-replicating life forms that we became.

3. If the phenotypic separation theory of sex is true, then…

- All organisms capable of reproducing through meiosis (including those in which meiosis is facultative) descend from a common pair of cells that used to be phenotypic (classically referred to as somatic), serving the reproduction of a genetic line of descent that used mitosis as a means of reproduction.

- If we could find the particular genes that were involved in multi-cellular activities in the organism subject to a phenotypic separation, and if modern meiotic single-cell organisms do descend from a phenotypic separation from this organism, then we may expect to find vestigial versions of genes that were useful to multi-cellular life even in single-cell meiotic organisms.

References

1. Holwerda G. (2013). *The Unbelievers*. Documentary.

2. Dawkins R. (1976). *The Selfish Gene*. Oxford University Press.

3. Dawkins R. (1982). *The Extended Phenotype*. Oxford University Press.

4. Kurzweil R. (2005). *The Singularity Is Near: When Humans Transcend Biology*. Viking.

5. Blackmore S. (2008). Memes and "temes." TED Lecture, https://www.ted.com/talks/susan_blackmore_on_memes_and_temes

6. Blackmore, S. (2009). Evolution's third replicator: Genes, memes, and now what? *New Scientist*, https://www.newscientist.com/article/mg20327191.500-evolutions-third-replicator-genes-memes-and-now-what/

7. Dennett, D. (1995). *Darwin's Dangerous Idea: Evolution and the Meanings of Life*. Simon & Schuster.

8. Dobzhansky T. (1973). Nothing in Biology Makes Sense Except in the Light of Evolution. *American Biology Teacher*.

9. Woese C.R. (1967). *The Genetic Code: The molecular basis for genetic expression*. Harper & Row.

10. Zurek W.H. (2003). Quantum Darwinism and Envariance. arXiv:quant-ph/0308163.

11. Forterre P. (2005). The two ages of the RNA world, and the transition to the DNA world: a story of viruses and cells. *Biochimie*, 87:793-803.

12. Blackmore S. (2002). The Evolution of Meme Machines. International Congress on Ontopsychology and Memetics, Milan. May 18-21. http://www.philo5.com/Textes-references/020727%20Blackmore%20-%20The%20Evolution%20of%20Meme%20Machines.htm

13. Blackmore S. (2000). *The Meme Machine*. Oxford University Press.

14. Gabora L. (1997). The Origin and Evolution of Culture and Creativity. *Journal of Memetics* - Evolutionary Models of Information Transmission, 1. http://cogprints.org/794/1/oecc.html

15. Gene, *Wikipedia*. Retrieved in December 2018. https://en.wikipedia.org/wiki/Gene

16. Leipe D.D., Araving L. and Koonin E.V. (1999). Did DNA replication evolve twice independently? *Nucleic Acids Research*, 27:3389-3401.

17. Weiner A.M., Maizels N. (1987). tRNA-like structures tag the 3' ends of genomic RNA molecules for replication: Implications for the origin of protein synthesis. *PNAS*, 84:7383-7387.

18. Greely H.T. (2016). *The End of Sex and the Future of Human Reproduction*. Harvard University Press.

Made in the USA
Middletown, DE
27 October 2020

22826236R00080